AN INTRODUCTION
TO DISCRETE SYSTEMS

AN INTRODUCTION TO DISCRETE SYSTEMS

KENNETH STEIGLITZ

Princeton University

with a foreword by M. E. Van Valkenburg

JOHN WILEY & SONS, INC.
New York
London
Sydney
Toronto

Library of Congress Cataloging in Publication Data

Steiglitz, Kenneth, 1939-
An introduction to discrete systems.

Includes bibliographies.
1. Digital filters.   2. Graph theory.   3. Electric networks.   I. Title.

TK7872.F5S74   621.3819'5   73-6820
ISBN 0-471-82097-0

Printed in the United States of America

10-9 8 7 6 5 4 3 2 1

To San

FOREWORD

Professor Steiglitz has invited me to write a few words about the origins of this textbook in relationship to its general educational objectives. Beginning in 1967, the Committee on Computers in Electrical Engineering (COSINE) of the National Academy of Engineering sponsored a series of task force meetings, each charged with examining a specific area of digital systems education and making recommendations for the development of course materials. The first of these, for which I was Chairman, examined a computer-oriented first course in electrical engineering, specifically in the linear circuits area. The meeting was held at Princeton University and involved about a dozen educators, one of whom was Kenneth Steiglitz.

The Task Force group recognized the now-accepted fact that the elementary circuits course provides an excellent medium for introducing the use of the digital computer through computer applications to numerous topics within linear circuits. It differentiated two ways in which this might be done: one was described as a *transition* approach and the other as an *integrated* approach.

The transition textbook was envisioned as containing more-or-less conventional topics in linear circuits, but presented with a computer orientation, with examples drawn from the application of numerical methods. Some of the topics that might be presented in an associated software laboratory or as homework assignments include: determining roots (Newton-Raphson method), numerical integration, the solution of linear algebraic equations (Gauss elimination method), the solution of ordinary differential equations (Runge-Kutta method), perhaps the solution of simple nonlinear differential equations, general network analysis using canned programs, the determination and plotting of magnitude and phase, and determination of residues.

The integrated approach to the teaching of a computer-oriented circuits course was envisioned as one that might evolve over a period of several years. Given the widespread availability of computers, there seems little doubt that the teaching of electrical engineering should undergo an evolution, with each traditional subject reexamined as to its relevance in a computer age, and each new subject examined to see where it might best fit into an educational program. In the emerging pedagogical approach, equations should be written in discrete form as difference equations, instead of in continuous form as differential equations. Indeed, equations should seldom be used, since principles should be stated directly in algorithmic form. Many concepts and related theorems that find use in computation will not survive, or will be useful only for checking. Thus a step-by-step restructuring of subject matter is envisioned.

A number of textbooks of the transition variety have already appeared, but this textbook appears to be the first using an integrated approach. Shortly after the meeting of the Task Force, Professor Steiglitz was persuaded to teach a course, based on an integrated approach, to first-term sophomores in electrical engineering at Princeton University. This course has been developed over the years with care, and several revisions of the organization and structure of the manuscript have taken place based on the experience with students. Having been associated with the course myself from time to time, I can report on some aspects that I have observed.

The notes have been well received by the sophomores. Since their only previous formal contact with the topics in circuits has come in the brief presentation in physics, they began their study of electrical engineering directly in terms of discrete signals and an algorithmic approach; they studied without apparent difficulty in comparison to a more conventional approach. This is in itself an interesting observation. I have discussed these subjects with faculty members at other institutions and teachers of courses in industrial laboratories in which the discrete case has been presented at the graduate level after the student has had extensive instruction in the continuous case. The universal reaction has been that it is a difficult subject for students at an advanced level. It appears that the discrete case is a natural one to the uninhibited mind, but special and somewhat mysterious after a thorough grounding in continuous concepts.

At Princeton, this first course has been followed in the second term by a conventional linear circuits course, sometimes facetiously described as the special limiting case of an infinite sampling rate. The similarity of concepts, language, and conclusions for the two cases was made throughout the second course. Students who have completed this two-term sequence have since continued through their last two years and on into graduate school or to industry. I believe that they have done so at an advantage; no disadvantages have become apparent from an arrangement of topics that many might think of as unconventional.

It has been a pleasure to talk with students who have returned from summer experiences in industry after their sophomore year. They found that the topics treated in the course gave them an immediate advantage in their job assignments.

To students about to embark on their study of the topics in this textbook, I promise a rewarding experience. Discrete-time systems and the processing of discrete-time signals will grow in importance in the future, making a knowledge of these subjects indispensable to the engineer. The basic ideas will become real through the artifice of the digital filter whose operation is easily visualized (or demonstrated with laboratory equipment) and readily implemented in the laboratory in either hardware or software form. Linear graphs used to study the relationships of quantities and the enumeration of things will be found useful in a wide variety of fields important to the engineer.

To faculty members who will guide the students through the pages of the textbook, I promise a lucid presentation of important subjects prepared by an authority on the topics presented, well stocked with interesting and challenging problems, and thoroughly tested in the classroom.

M. E. Van Valkenburg
Princeton University

PREFACE

This book evolved over the past four years as notes for a one-semester, first course in electrical engineering at Princeton University. The premises behind the choice of material are that the discrete-time, digital device has become the fundamental building block of engineering systems; that first- or second-year students are often already prejudiced towards thinking in terms of continuous-time systems; and that this prejudice should be counteracted as soon as possible. The text is designed, therefore, so that it can precede a conventional circuit theory course, and the last chapter provides an introduction to circuit theory that can serve as a transition to such a course. This sequence has the advantage of introducing discrete-time linear system theory, which makes no use of differential equations or δ-functions, earlier in the curriculum than the mathematically more advanced continuous-time theory. This reinforces the student's eventual appreciation of the common fundamental notions of both theories. This arrangement has the additional advantage of providing a first course in electrical engineering that focuses on the digital computer, thus providing material that is relevant to computer science students.

The book also can be used in conjunction with a textbook in circuit theory for a "hybrid" course in discrete- and continuous-time systems. This idea—of using a pair of texts as coroutines—offers important advantages: variety of motivation, strengthening of appreciation for generality, and great flexibility for the instructor. Other possibilities are using either Part I or II separately for a one-quarter course, or omitting Chapters 8 to 10 in a faster one-quarter course for students with a background in continuous systems.

Most of the material is new to the undergraduate curriculum, and the working engineer may wish to use the text for self-study. His familiarity with the Laplace transform and computer programming should enable him to use this material as an introduction to current literature in the fields of digital signal processing and networks for communication, transmission, and transportation.

I am indebted to many people for contributions to this project. First, I thank Professor Mac E. Van Valkenburg, who first pointed out to me the need for a computer-oriented first course, and who helped and encouraged me in many ways. Without him, this book would never have been written. Dr. Godfrey Winham read many early versions of the text with great care, and his critical comments improved the result considerably. His program for the FFT is given in Chapter 6. The treatment of the Ford and Fulkerson labeling algorithm is based on a talk by Dr. Bill Rothfarb, I am indebted to Professors T. L. Booth, J. L. Bruno, S. W. Director, E. I. Jury, T. Pavlidis, and R. J. Smith for many constructive comments on the manuscript; and to G. A. Davenport, D. C. Ford, and others at Wiley for their splendid cooperation. Finally, I thank the students of Engineering 283. I hope they learned as much from me as I learned from them.

Kenneth Steiglitz

A NOTE TO THE READER

This book is an introduction to the use of the digital computer in engineering. The fundamental idea is that of the step-by-step procedure: the algorithm. In Part I, algorithms for processing signals are studied: in Part II, algorithms for solving problems associated with systems (in a general sense) are taken up. An attempt has been made to keep the material fundamental and general and, thus, to

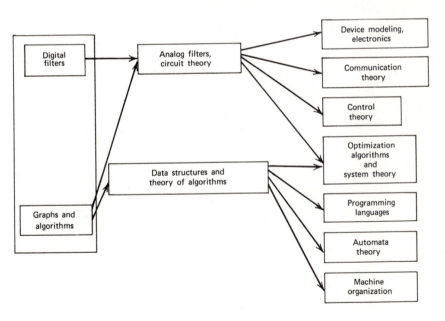

provide a foundation for future courses, including (but not restricted to) a conventional course in circuit analysis.

Above is a block diagram with the subjects of this book, together with a rough (and somewhat arbitrary) indication of how they lead to other subjects in engineering and computer science.

Although some of the exercises are routine and are meant to test your comprehension of material, many are intended to lead deeper into the material and to challenge your wit. Especially difficult exercises are indicated by an asterisk. Answers and hints to many of the exercises are given in the Appendix.

It is assumed that you have had the equivalent of an introductory course in computer programming and are familiar with FORTRAN.

The programs given in the text have been written in WATFIV. Some of the exercises ask you to write programs; these programs can, of course, be written in any language you find convenient.

Finally, a word about mathematical requirements. The relevant mathematics for Part I is complex numbers; that for Part II is graph theory. Both subjects are introduced and explained in the text (Chapters 1 and 7). If you are not familiar with operations with complex numbers, you will find that some practice is required to gain the facility required later on. If you find yourself with free time before a course in this material, some study of complex numbers will prove helpful.

                                                                              K.S.

CONTENTS

GRAPHS AND ALGORITHMS

DIGITAL FILTERS

1.
COMPLEX NUMBERS

1.1 INTRODUCTION

This book has two parts, the first dealing with linear discrete-time systems, or digital filters; and the second with graphs and algorithms for solving some problems that can be formulated in terms of graphs. These subjects were chosen for an introductory text in engineering for a number of reasons. First, these two areas of knowledge provide a basis for formulating and solving a large variety of engineering problems dealing with signal processing, system design, and resource allocation. Second, this material provides an introduction to the use of computers in several ways: as a processor of signals, as a means of computing functions, and as a means of storing and manipulating descriptions of complex systems. Another reason for choosing this material, one that hardly needs mention, is that the author has used these techniques in the practice of engineering.

Part I deals with the digital processing of signals that bear information. The reason for restricting ourselves to *digital* processing is that with today's electronic technology it has become advantageous to keep information in *digital form*; that is, in the form of numbers. One reason for this is clear: information stored as numbers is safer, because it is less likely to be eroded by the influence of the physical surroundings. As an example, suppose we have a piece of information concerning the altitude of an aircraft. Let us say that the altitude at a certain time is 23,500 ft. How can we keep this information, so that it is available when we need it and, further, so that we can retrieve it quickly and accurately? Twenty years ago the answer might have been: as the magnitude of the voltage on a capacitor in an electrical circuit. As a voltage, however, the piece of information is susceptible to a number of disturbing influences. First, the voltage on a capacitor can change as a result of the electrons on the plates leaking away.

Second, the accuracy with which we can retrieve the information is limited by the inevitable presence of electrical noise. Third, variations in temperature and humidity can affect the voltage.

Today, the answer might go something like this: in the binary number system, 23,500 is

$$101101111001100 \qquad (1.1.1)$$

This represents 15 bits, or yes-no pieces of information. Build 15 devices, each of which is either *on* (indicating 1) or *off* (indicating 0). Line these up and turn on the appropriate ones. Thus, each box in Fig. 1.1.1. represents a bit, and the row of 15 boxes in the states shown represents the number 23,500.

The success of this scheme depends, of course, on our ability to build boxes that are in one of two clearly defined states. We require that each box can be set into either state easily and rapidly, that the box will stay in a given state indefinitely, and that the state of a given box can be read at any time, quickly and without changing the state. Such boxes, which may be called *binary storage devices,* are easy and cheap to build. They are at the heart of computer technology. Core storage, disks, digital tapes, and registers, are all comprised of binary storage devices. The reason for the success of this technology depends strongly on the fact that we require these boxes to have only *two* identifiable states, while we require our capacitor (which might be termed an *analog* storage device) to have a very large number of identifiable states, one for each possible voltage. Quite simply, we have made the task much easier, so that we can do a much better job.

Information stored as numbers can be combined easily with other information also stored that way. In addition, it can be transmitted over long distances under adverse circumstances without degradation. This is the basis of pulse-code modulation (PCM), which has revolutionized the communication industry. In short, the digital form is ideal for the purposes of system design and, for this reason, we shall deal with information in digital form in this introduction to signal processing.

The starting point for our study will be the *linear processing* of digital signals, a relatively new field, which is called digital filtering. The reason for this choice is simple: by restricting ourselves to *linear* processes, a great deal of insight can be gained into the properties of signals in general. In

FIG. 1.1.1   Representation of a binary number with binary storage devices.

particular, the introduction of *frequency domain* concepts, in which a signal is thought of as being composed of sums of sinusoidal components, allows us to design complex and useful processors. The properties of linear digital signal processors have profound analogies with properties of linear systems in general, so that the techniques developed will be directly applicable to other areas, such as linear circuit theory. An essential prerequisite for this study is complex numbers, the subject of the rest of this chapter.

1.2 THE FIELD OF COMPLEX NUMBERS

We are all familiar with real numbers. Not only can we add, subtract, multiply, and divide them, but these operations correspond to intuitive concepts learned from observation of the physical world. If asked to divide the real number a by the real number b, we can ask ourselves: How many times does b go into a? If a is the length of a road, and b is the length of a piece of chain, then a/b is the number of times the chain must be laid end-to-end to traverse the road. Our experience with length supports our understanding of what a real number is.

On the other hand, complex numbers may seem more abstract and remote from our everyday experience. If we were asked to divide the complex number a by the complex number b, we would in all likelihood be able to find the correct answer. But we might have difficulty checking the reasonableness of our answer. We would have little "feeling" for what the result might be. This is unfortunate and a situation that we hope to change here since complex numbers are very important in describing the operation of digital filters.

The set of real numbers together with the operations discussed above comprise a mathematical structure called a *field*. What this means, among other things, is that the sum and product of two real numbers are always defined, as well as the inverses of these operations (except for division by zero). The set of complex numbers also comprise a field, and the addition and multiplication of complex numbers, as well as their subtraction and division, are always defined (except, again, in the case of division by zero). This section is devoted to the definitions of these operations on complex numbers.

A complex number z is thought of as having two parts; a real part, say x, and an imaginary part, say y: both are real numbers. We express this by writing

$$z = x + jy \qquad (1.2.1)$$

Think of this as meaning that the complex number z is a point in a plane (called the complex plane), with abscissa x and ordinate y, as shown in Fig. 1.2.1.

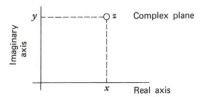

FIG. 1.2.1   A number z in the complex plane.

The abscissa is called the *real axis*, and the ordinate is called the *imaginary axis*. Thus a complex number is really just a pair of real numbers. The special properties of complex numbers result from the way in which we define addition and multiplication.

Given two complex numbers, say z_1 and z_2, we define their sum, $z_1 + z_2$, as the complex number with a real part that is the sum of the real parts of z_1 and z_2; and an imaginary part that is the sum of the imaginary parts of z_1 and z_2. Symbolically, if

$$z_1 = x_1 + jy_1 \qquad (1.2.2)$$

and

$$z_2 = x_2 + jy_2 \qquad (1.2.3)$$

then we define

$$z_1 + z_2 = (x_1 + x_2) + j(y_1 + y_2) \qquad (1.2.4)$$

The process of addition is illustrated in Fig. 1.2.2. We see from this figure that the sum of two complex numbers can be obtained in the same way that we obtain the sum of two vectors. If we think of a complex number as a vector from the origin in the complex plane to the point in the complex plane with the appropriate coordinates, then the sum of two complex numbers can be obtained by moving the origin of one of the vectors so that it coincides with the tip of the other. If we do this in the two possible ways, first moving one and then the other, we obtain a parallelogram, and so the process of vector addition is sometimes called the *parallelogram law*. Displacements, forces, velocities, and all other vector quantities can be added in this way.

In order to define the product of two complex numbers, we shall introduce another way of representing them. Instead of thinking of a complex

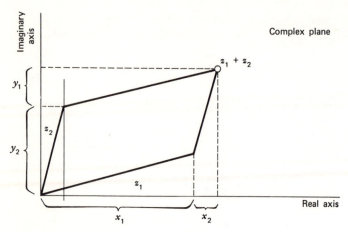

FIG. 1.2.2 Addition of complex numbers.

number as a real and imaginary part, we can also think of it as a magnitude and a direction. The *magnitude* of z, the distance from the origin to the point in the complex plane, is sometimes called the *modulus*, and is denoted by $|z|$. The direction, measured by the angle between the real line and the vector to the point in the complex plane, is called either the *angle* or the *argument* of the complex number, and is denoted by Arg z. Figure 1.2.3 illustrates the definition of the magnitude, R, and angle, θ.

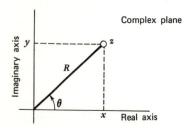

FIG. 1.2.3 Polar representation of a complex number.

From this diagram, we can determine the relations between the two sets of coordinates. The magnitude and angle can be written in terms of the real and imaginary parts as

$$R = \sqrt{x^2 + y^2} \qquad (1.2.5)$$

$$\theta = \arctan \frac{y}{x} \qquad (1.2.6)$$

Or, we can write the real and imaginary parts in terms of the magnitude and angle as

$$x = R \cos \theta \qquad (1.2.7)$$

$$y = R \sin \theta \qquad (1.2.8)$$

We can now define the product of two complex numbers as follows:

Let z_1 and z_2 be complex numbers with magnitudes R_1 and R_2, respectively; and with angles θ_1 and θ_2, respectively. Then the product $z_1 z_2$ is that complex number with magnitude $R_1 R_2$ and angle $\theta_1 + \theta_2$.

A number of observations concerning this definition are in order. First, we would like to think of real numbers as special cases of complex numbers in the sense that the imaginary parts of real numbers are zero. The above definition of multiplication is consistent with this view. To check this, consider two real numbers that are positive. They can be thought of as complex numbers with magnitudes equal to the real numbers and angles that are zero. On multiplication as complex numbers, the result still has angle zero, and has the correct magnitude: the product of the two real numbers. If one of the real numbers is positive and one negative, the angle of the first is zero, and the angle of the second is π. On multiplication, the result has angle π, and a magnitude that is the product of the magnitudes of the two real numbers. This checks, since the result must be negative. Similar argument checks the case when both real numbers are negative, since the result has angle 2π, which is equivalent to having angle 0.

A second observation concerns the special complex number with real part zero and imaginary part equal to 1. This can be thought of as having magnitude 1 and angle $\pi/2$. Denote this complex number by j, according to the notation used in (1.2.2). Now let us calculate $j \cdot j$. Its magnitude is 1, by the definition of multiplication; and similarly, its angle is π. Hence

$$j \cdot j = -1 \qquad (1.2.9)$$

This confirms the interpretation that the imaginary part of a complex number represents a multiple of the square root of -1. This also accounts for the usefulness of complex numbers in representing the roots of polynomials. For example, the polynomial

$$w^2 + 1 = 0 \qquad (1.2.10)$$

has no solutions if w is confined to the field of real numbers. If we allow w to be an element of the complex field, it has two solutions: $\pm j$. The fundamental theorem of algebra states, in fact, that every polynomial of degree n has precisely n roots in the complex field. This result is the primary reason why complex numbers are useful.

The operations of subtraction and division are defined simply as the inverses of the operations of addition and multiplication, respectively. Thus, $z_1 - z_2$ is obtained simply by subtracting the real and imaginary parts of z_1 and z_2 in turn. To divide z_1 by z_2, we simply divide the magnitude of z_1 by the magnitude of z_2; and subtract the angle of z_2 from the angle of z_1.

We now shall consider the effect of multiplication on the real and imaginary parts of the complex numbers involved. Let us introduce the following further notation: if z has magnitude R and angle θ, we write

$$z = R \angle \theta \qquad (1.2.11)$$

(Read as "R at an angle θ.") Now let

$$z_1 = R_1 \angle \theta_1 = x_1 + jy_1 \qquad (1.2.12)$$

and

$$z_2 = R_2 \angle \theta_2 = x_2 + jy_2 \qquad (1.2.13)$$

Then by our definition of multiplication

$$z_1 z_2 = R_1 R_2 \angle \theta_1 + \theta_2 \qquad (1.2.14)$$

Hence, the real part of the product is

$$R_1 R_2 \cos(\theta_1 + \theta_2)$$

If we use the trigonometric identity

$$\cos(\theta_1 + \theta_2) = \cos\theta_1 \cos\theta_2 - \sin\theta_1 \sin\theta_2 \qquad (1.2.15)$$

the real part can be written

$$R_1 \cos\theta_1 R_2 \cos\theta_2 - R_1 \sin\theta_1 R_2 \sin\theta_2 \qquad (1.2.16)$$

We recognize here the real and imaginary parts of z_1 and z_2, and we can rewrite the real part of the product as

$$x_1 x_2 - y_1 y_2 \qquad (1.2.17)$$

Similarly, the imaginary part of the product is

$$R_1 R_2 \sin(\theta_1 + \theta_2) = R_1 \sin\theta_1 R_2 \cos\theta_2 + R_1 \cos\theta_1 R_2 \sin\theta_2 = y_1 x_2 + x_1 y_2$$
$$(1.2.18)$$

Hence, the product is

$$z_1 z_2 = (x_1 x_2 - y_1 y_2) + j(y_1 x_2 + x_1 y_2) \qquad (1.2.19)$$

This is precisely what results if we multiply (1.2.12) and (1.2.13), interpreting j^2 as -1:

$$(x_1 + jy_1)(x_2 + jy_2) = x_1 x_2 + jy_1 x_2 + jx_1 y_2 + j^2 y_1 y_2$$
$$= (x_1 x_2 - y_1 y_2) + j(y_1 x_2 + x_1 y_2) \qquad (1.2.20)$$

Thus, it is permissible to multiply complex numbers written in the form $x + jy$, provided that we interpret j^2 as -1.

One final bit of terminology: if z is written as $x + jy$, we say z is in *Cartesian* form; if it is written as $R\angle\theta$, we say z is in *polar* form. These terms follow, of course, from the usual names of the corresponding coordinate systems.

Example Problem

If y and z are two complex numbers, and $yz = 0$, show that either $y = 0$ or $z = 0$.

Solution. From the definition of multiplication

$$|yz| = |y||z|$$

Since $|yz| = 0$, this implies that either $|y| = 0$ or $|z| = 0$, which implies that either $y = 0$ or $z = 0$.

Example Problem

Express the complex number $(3 + j5)^4$ in the form $a + jb$.

First Solution. From (1.2.20) we know that we can use the usual rules of algebra, treating j^2 as -1. Thus

$$(3 + j5)^2 = 9 + j30 - 25 = -16 + j30$$

Squaring this, we get

$$(3 + j5)^4 = 256 - j960 - 900 = -644 - j960$$

Second Solution. If we go back to the definition of multiplication, we must first put $(3 + j5)$ in polar form. If we write

$$(3 + j5) = R\angle\theta$$

then

$$R = \sqrt{3^2 + 5^2} = \sqrt{34}$$

$$\theta = \arctan \frac{5}{3} = \arctan 1.667 = 1.03038$$

Now we can find

$$(3+j5)^4 = R^4 \underline{/4\theta}$$

from the definition of multiplication. Hence, the real part of $(3 + j5)^4$ is, using (1.2.7),

$$R^4 \cos 4\theta = 1156 \cos 4.12152 = -644.01$$

Using (1.2.8), we have similarly

$$R^4 \sin 4\theta = 1156 \sin 4.12152 = -959.99$$

Third Solution. If a computer with a FORTRAN compiler is available, we can run the following program

```
COMPLEX Z
Z=(3.,5.)
Z=Z**4
WRITE(6,1)Z
1 FORMAT(' Z=',2F12.5)
END

Z=   -644.00000   -960.00000
```

Exercise 1.2.1

Let

$$z_1 = x_1 + jy_1 = R_1 \underline{/\theta_1} \tag{1.2.21}$$

$$z_2 = x_2 + jy_2 = R_2 \underline{/\theta_2} \tag{1.2.22}$$

From the definition of division:

$$\frac{z_1}{z_2} = \frac{R_1}{R_2} \underline{/\theta_1 - \theta_2} \tag{1.2.23}$$

show that

$$\frac{z_1}{z_2} = \frac{x_1 x_2 + y_1 y_2}{x_2^2 + y_2^2} + j \frac{y_1 x_2 - x_1 y_2}{x_2^2 + y_2^2} \tag{1.2.24}$$

Show also how this result can be obtained by direct algebraic manipulation of

$$\frac{x_1 + jy_1}{x_2 + jy_2} \tag{1.2.25}$$

Exercise 1.2.2

What effect does taking the square root of a complex number have on its magnitude and angle? What effect does taking the nth root? Raising to the nth power?

Exercise 1.2.3

The *complex conjugate* of $z = a + jb$ is defined as $z^* = a - jb$. What is the complex conjugate of z in polar form? Express the real part, imaginary part, and the magnitude of z, in terms of z and z^*.

Exercise 1.2.4

Plot in the complex plane the 5th roots of 1, of -1, and of j.

Exercise 1.2.5

Prove de Moivre's theorem:

$$(\cos \theta + j \sin \theta)^n = \cos n\theta + j \sin n\theta \qquad (1.2.26)$$

for integer n.

Exercise 1.2.6

The full definition of a field is as follows: A set of elements F is called a *field* if:

1. Each pair of elements of F determines uniquely a sum $a + b$ in F, such that

   (a) $\quad a+b=b+a \qquad$ (commutative law)
   (b) $\quad a+(b+c) = (a+b) +c \quad$ (associative law)

2. Each pair of elements of F determines uniquely a product ab in F such that

   (a) $\quad ab=ba \qquad$ (commutative law)
   (b) $\quad a(bc) = (ab)c \quad$ (associative law)

3. $a(b+c) = ab+ac$ (distributive law).

4. F contains distinct elements 0 and 1 that act as identities for addition and multiplication, respectively.

5. For each a in F the equation $a + x = 0$ has in F a solution, denoted by $-a$ (existence of an inverse to addition).

6. For each $a \neq 0$ in F the equation $ax = 1$ has in F a solution, denoted by a^{-1} (existence of an inverse to multiplication).
(See the suggestions for further reading, numbers 5 to 7.)

Verify these axioms for the set of complex numbers, assuming the definitions of addition and multiplication given above.

Exercise 1.2.7

Suppose we define multiplication for complex numbers as follows:

$$(x_1 + jy_1)(x_2 + jy_2) = x_1 x_2 + jy_1 y_2$$

Would the set of complex numbers with multiplication defined in this way be a field? Explain your answer.

Exercise 1.2.8

Let F be the set of all real numbers of the form $a + b\sqrt{2}$, where a and b are rational numbers. Show that with the ordinary definitions of sum and product for real numbers, F is a field.

Exercise 1.2.9*

On many computers, the time required to perform real addition is very small compared with the time to perform real multiplication. In order to

---

* Attributed to G. Golub by R. C. Singleton in "An Algorithm for Computing the Mixed Radix Fast Fourier Transform," R. C. Singleton, *IEEE Transactions on Audio and Electroacoustics*, vol. AU-17, pp. 93–103, 1969.

calculate the product of two complex numbers, say $z = a + jb$ and $w = c + jd$, it would seem at first to require four real multiplications, one for each product in the formula

$$wz = (ac - bd) + j(ad + bc)$$

Show that it is possible to compute wz with 3 real multiplications, 5 real additions, and 2 negation operations (taking the negative of a real number).

Exercise 1.2.10

Show that for any two complex numbers w and z,

$$|w + z| \le |w| + |z|$$

Find the conditions under which .equality holds in this inequality. (This relation is known as the *triangle inequality*.)

Exercise 1.2.11

Show that for any two complex numbers w and z

$$|w + z| \ge |\,|w| - |z|\,|$$

Find the conditions under which equality holds in this inequality.

Exercise 1.2.12

(a) Show that if $P(z)$ is a polynomial in the complex variable z having all its coefficients real, then $P(z^*) = [P(z)]^*$, where $()^*$ is the complex conjugate (see Ex. 1.2.3).

(b) A polynomial $P(z)$ is said to have *root symmetry* if $P(z) = 0$ implies $P(z^*) = 0$. Show that a polynomial has root symmetry if it has real coefficients. Is the converse true?

Exercise 1.2.13

Combine the binomial theorem with de Moivre's theorem (1.2.26) to obtain an expression for $\cos n\theta$ in terms of $\sin \theta$ and $\cos \theta$.

1.3 EULER'S FORMULA

Having defined arithmetic operations for complex numbers, we could proceed to show that the usual theorems of algebra and calculus hold in this more general setting. We shall, however, assume that you are familiar with this mathematics and, instead, we shall concentrate on applications to digital filters.

Power series will play an especially important role in what we want to do. The convergence of a series of complex numbers is treated in the same way as the convergence of a series of real numbers; for example, we can show that the infinite series

$$1 + z + z^2 + z^3 + \ldots \tag{1.3.1}$$

converges to the function

$$\frac{1}{1-z} \tag{1.3.2}$$

provided that the complex number z has modulus less than 1. This is analogous to the real series of the same form, which converges to the same function when the real number involved is less than 1 in magnitude. The part of the complex plane where the series (1.3.1) converges is called the *region of convergence*. Here it consists of the region inside the circle of radius 1 centered at the origin, shown as the shaded region in Fig. 1.3.1.

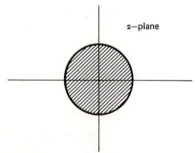

FIG. 1.3.1   The region of convergence of the series of Eq. 1.3.1.

The function e^z can be defined when z is a complex number by the infinite series

$$1 + \frac{z}{1!} + \frac{z^2}{2!} + \frac{z^3}{3!} + \ldots \tag{1.3.3}$$

This series converges for every complex number z, and coincides with the usual definition when z is real. Now consider the case when z is pure imaginary; that is, when z has real part zero. If we write z as $j\theta$ in this case, the power series (1.3.3) becomes

$$1 + \frac{j\theta}{1!} + \frac{(j\theta)^2}{2!} + \frac{(j\theta)^3}{3!} + \ldots \qquad (1.3.4)$$

The terms can be grouped into two parts, according to whether they are real or complex, as follows:

$$\left(1 - \frac{\theta^2}{2!} + \frac{\theta^4}{4!} - \ldots\right) + j\left(\frac{\theta}{1!} - \frac{\theta^3}{3!} + \ldots\right) \qquad (1.3.5)$$

The real part will be recognized as the power series for the cosine function of a real variable θ, and the imaginary as the series for $\sin \theta$. We have therefore shown in a formal way that

$$e^{j\theta} = \cos \theta + j \sin \theta \qquad (1.3.6)$$

This formula is called *Euler's formula,* and is fundamental in the rest of our discussion of digital signal processing. Notice that this formula is consistent with de Moivre's theorem, (1.2.26), which translates into

$$(e^{j\theta})^n = e^{jn\theta} \qquad (1.3.7)$$

Now any complex number that in polar form is $R \underline{/\theta}$ can be written as

$$R \cos \theta + jR \sin \theta = R (\cos \theta + j \sin \theta) \qquad (1.3.8)$$

which, by Euler's formula, becomes simply

$$Re^{j\theta} \qquad (1.3.9)$$

Thus, $e^{j\theta}$ can be thought of as the "angle part" of any complex number. Its modulus is 1, as can be seen directly from Euler's formula. In fact, we can read $Re^{j\theta}$ as "R at an angle θ," and $e^{j\theta}$ can be thought of simply as $1\underline{/\theta}$.

The function $e^{j\theta}$, like the trigonometric functions, is not changed by adding or subtracting an integer multiple of 2π to θ. In other words, $e^{j\theta}$ is a periodic function of θ, with period equal to 2π. This is a consequence of the fact that the angle of a complex number is defined ambiguously; we did not, when we defined the angle of a complex number, specify how to decide between, for example, $-\pi/4$ and $7\pi/4$. This will not cause any difficulty in what we shall do. Unless stated otherwise, we shall arbitrarily take θ so that it is between $-\pi$ and π.

It can be shown from the defining power series (1.3.3) that the exponential function of a complex variable obeys the addition formula that we expect it to. That is, if z and w are complex numbers, then

$$e^{z+w} = e^z e^w \tag{1.3.10}$$

In particular,

$$e^{a+jb} = e^a e^{jb} = e^a (\cos b + j \sin b) \tag{1.3.11}$$

which puts the exponential function of a complex variable in terms of real functions.

Example Problem

(a) Interpret geometrically the difference $z = w - u$ between two complex numbers.

(b) Find an expression for $|z|$ in terms of $|w|$, $|u|$, and the angle between w and u.

Solution. (a) The complex number z is that number which must be added to u in order to get w. That is, $u + z = w$. If we depict w as the sum of the vectors u and z, as in the figure below, we see that $z = w - u$ *is the complex number represented by the vector drawn with starting point at u and end point at w.*

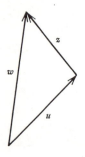

(b) Let us write

$$z = R_1 e^{j\theta_1} \qquad w = R_2 e^{j\theta_2} \qquad u = R_3 e^{j\theta_3}$$

Then the relation $z = w - u$ becomes

$$R_1 e^{j\theta_1} = R_2 e^{j\theta_2} - R_3 e^{j\theta_3}$$

This is an equation between complex numbers and, hence, we can break it down into two real equations by equating the real parts and the imaginary parts of both sides. Using Euler's formula, we obtain

$$R_1 \cos \theta_1 = R_2 \cos \theta_2 - R_3 \cos \theta_3$$
$$R_1 \sin \theta_1 = R_2 \sin \theta_2 - R_3 \sin \theta_3$$

We want to solve for R_1 in terms of the known quantities on the right-hand sides of these equations. This can be accomplished by squaring both equations, adding, and using the fact that $\cos^2 \theta_1 + \sin^2 \theta_1 = 1$:

$$R_1{}^2 = R_2{}^2 + R_3{}^2 - 2R_2R_3 (\cos \theta_2 \cos \theta_3 + \sin \theta_2 \sin \theta_3)$$

Using the further trigonometric identity

$$\cos \theta_2 \cos \theta_3 + \sin \theta_2 \sin \theta_3 = \cos (\theta_2 - \theta_3)$$

this becomes

$$R_1{}^2 = R_2{}^2 + R_3{}^2 - 2R_2R_3 \cos (\theta_2 - \theta_3)$$

which is the desired result, since $\theta_2 - \theta_3$ is the angle between w and u. (This equation is known as the *law of cosines.*)

Example Problem

(a) Prove that for all complex z (except $z = 1$)

$$1 + z + z^2 + \ldots + z^k = \frac{1 - z^{k+1}}{1 - z} \qquad k = 1, 2, \ldots \qquad (*)$$

(b) From this prove that if $|z| < 1$

$$1 + z + z^2 + \ldots = \lim_{k \to \infty} (1 + z + z^2 + \ldots + z^k) = \frac{1}{1 - z}$$

Solution. (a) We shall use the principle of *mathematical induction.* That is, we shall show

(1) That $(*)$ is true for $k = 1$.

(2) That if $(*)$ is true for k, it is true for $k + 1$.

When $k = 1$ the right-hand side of $(*)$ becomes

$$\frac{1 - z^2}{1 - z} = \frac{(1 - z)(1 + z)}{1 - z} = 1 + z$$

which checks with the left-hand side. Thus, ($*$) is true for $k = 1$. Assume now that ($*$) holds for k, and consider the case $k + 1$:

$$1+z+z^2+\ldots+z^{k+1} = (1+z+z^2+\ldots+z^k) + z^{k+1}$$

Using ($*$) for the expression in parentheses, this becomes

$$1+z+z^2+\ldots+z^{k+1} = \frac{1-z^{k+1}}{1-z} + z^{k+1} = \frac{1-z^{k+2}}{1-z}$$

which shows that ($*$) is true for $k + 1$. This completes the inductive proof, and shows that ($*$) holds for all positive integers k.

(b) To show that a sequence of complex numbers, say f_k, approaches the complex number f, we must show that

$$\lim_{k \to \infty} |f_k - f| = 0$$

(This is the definition of convergence of a sequence of complex numbers, and uses as a criterion the convergence of a sequence of real numbers: the sequence $|f_k - f|$.) In our case

$$f_k = 1 + z + z^2 + \ldots + z^k$$

and

$$f = \frac{1}{1-z}$$

Using ($*$), we can calculate

$$|f_k - f| = \left| \frac{1-z^{k+1}}{1-z} - \frac{1}{1-z} \right| = \left| \frac{-z^{k+1}}{1-z} \right|$$

It can be shown (see Exercise 1.3.11) that the magnitude of a ratio of complex numbers is equal to the ratio of magnitudes, and similarly that the magnitude of a product of complex numbers is equal to the product of magnitudes. Using this, we get

$$|f_k - f| = \frac{|-z^{k+1}|}{|1-z|} = \frac{|z|^{k+1}}{|1-z|}$$

Assuming that $|z| = R < 1$,

$$\lim_{k \to \infty} |f_k - f| = \lim_{k \to \infty} \frac{R^{k+1}}{|1-z|} = \frac{1}{|1-z|} \lim_{k \to \infty} R^{k+1} = 0$$

which is what we wished to prove.

Exercise 1.3.1

Express the sine and cosine functions of real variables in terms of the exponential function of a complex variable.

Exercise 1.3.2

Find the modulus and angle of the function

$$F = 1 + z \tag{1.3.12}$$

when $z = e^{j\theta}$, as functions of θ. Sketch in the complex plane the locus of F as θ varies from 0 to 2π.

Exercise 1.3.3

Find the modulus and angle of the function

$$F = \frac{1}{1-z} \tag{1.3.13}$$

when z is pure imaginary. Sketch F in the complex plane as z moves up the imaginary axis from $-j\infty$ to $+j\infty$.

Exercise 1.3.4

Let $y = \tan x$. Show that

$$e^{2jx} = \frac{1+jy}{1-jy} \tag{1.3.14}$$

Exercise 1.3.5

Put the following complex numbers in polar form:

    (a) $3 + j4$ (d) $e^3 + j4$

    (b) $7 + j7$ (e) $e^{(e^j)}$

    (c) $3 - j4$

Exercise 1.3.6 (computer experiment)

The purpose of this experiment is to test Euler's formula using the defining power series and complex arithmetic. Write a FORTRAN program that performs the following calculations:

(a) Read in a complex number z. This can be done by reading in the real and imaginary parts.

(b) Having declared the variable z complex in a FORTRAN declaration statement, calculate the series (1.3.3) using complex arithmetic. Use as a criterion of convergence the magnitude of each additional term; that is, test the magnitude of each term and stop when it falls below some small prescribed number. *Hint:* do NOT calculate $n!$ for each term. Instead, use the following loop: having computed the $(n - 1)$st term of the series, the nth term is obtained by multiplication by z/n.

(c) Compare the answer with e^z as computed using the built-in complex exponential function CEXP.

(d) Compare the real and imaginary parts of the answer with those computed using the built-in real EXP, SIN, and COS functions in (1.3.11).

(e) Include in your printout the number of terms the series required for convergence.

Run the program with data for several values of z, including 0, $j\pi/2$, $j\pi$, $j3\pi/2$, and $j2\pi$.

Exercise 1.3.7

Find out how CEXP, EXP, SIN, and COS are computed by the FORTRAN built-in library subprograms that you use.

Exercise 1.3.8

Show from the defining power series that (1.3.10) is true.

Exercse 1.3.9

Describe with a rough plot the motion of the point z^t as t increases from 0 to ∞ for the following values of z:

(a) z real, positive, less than 1

(b) z real, positive, greater than 1

(c) z real, negative, greater than -1

(d) z real, negative, less than -1

(e) $z = -1$

(f) $z = j$

(g) $z = -j$

(h) $z = 2j$

(*Hint:* the answer is not always unique.)

Exercise 1.3.10

Using Euler's formula, prove the following identity for the sum of a constant and N harmonically related cosine waves:

$$1 + 2\,(\cos x + \cos 2x + \ldots + \cos Nx) = \frac{\sin\,(N + \tfrac{1}{2})\,x}{\sin\,x/2} \qquad (1.3.15)$$

Prove the relation again using mathematical induction.

Exercise 1.3.11

(a) Show that for any complex numbers z and w, and for all integers p (positive, negative, and zero),

$$|z^p| = |z|^p$$
$$|zw| = |z| \cdot |w|$$

(b) Use this result to calculate

$$\left| \frac{1}{(3 + j4)\,(6 + j8)} \right|$$

Exercise 1.3.12

Show that the curve in the z-plane defined by the equation

$$|z - a| = R$$

is a circle with center a and radius R.

Exercise 1.3.13 (computer experiment)

Repeat Exercise 1.3.6, testing, instead of Euler's formula, the following closed form formula for a geometric series:

$$1+z+z^2+z^3+\ldots=\frac{1}{1-z}$$

Exercise 1.3.14

Show that $(z - 1)/(z + 1)$ is on the imaginary axis if and only if z is on the unit circle.

Exercise 1.3.15

Compute the real and imaginary parts of the complex number $(1 + j)^{100}$.

Further Reading

The following books will provide the reader with further discussion and exercises related to complex numbers.

1. *College Algebra,* Charles H. Lehmann, Wiley, New York, 1962. Chapter 8.

2. *Modern Algebra with Trigonometry,* John T. Moore, Macmillan, New York, 1969.

3. *Calculus and Analytic Geometry,* G. B. Thomas, Jr., Addison-Wesley, Reading, Mass., 1968 (fourth edition). Chapter 19.

4. *Modern Algebra and Trigonometry,* E. P. Vance, Addison-Wesley, Reading, Mass., 1962. Chapter 16.

The second book also includes a discussion of mathematical induction when de Moivre's theorem is discussed. For a more abstract treatment of complex numbers, see references 5–7 below.

5. *Introduction to Higher Algebra,* A. Mostowski and M. Stark, (translated from the Polish), Macmillan, New York, 1964. Chapter III.

6. *A Survey of Modern Algebra,* G. B. Birkhoff and S. Maclane, Macmillan, New York, 1941.

7. *Algebra—A Modern Introduction,* John L. Kelley, Van Nostrand, Princeton, New Jersey, 1965. Chapter 4.

Reference 7 contains an interesting discussion of the possibility of defining hypercomplex numbers—numbers with more than two components. Finally, see the historical notes in reference 8.

8. *Algebra—An Elementary Text-Book,* G. Chrystal, Dover, New York, 1961, (republication of an original edition of 1904). Chapter XII.

2.
DIGITAL SIGNALS
AND PHASORS

2.1 DIGITAL AND ANALOG SIGNALS

We are familiar with the notion of a signal as being information borne by some medium. When we speak of "radio signals," we mean information carried by electromagnetic waves of a particular frequency. When we speak of "visual signals," we mean information carried by light waves, which is electromagnetic radiation in a different part of the spectrum. Or we speak of "sound signals," for which the medium is compression waves in the air. The two aspects of a signal are the *information* and the *medium*. It often happens that the same information is transferred from one medium to another; this happens, for example, when we record sound waves with a microphone onto magnetic tape. Here, the same information is first carried by the sound waves, then by the physical displacement of a diaphragm in the microphone, then by electrical waves produced in the microphone, and finally by a magnetic field in the iron particles on the magnetic tape recording. Much of our present technology is concerned with the transfer of signals from one medium to another.

Signals generally vary with time, and this is the important way in which information is manifested. It is a dull signal indeed that merely stays constant: such a constant signal can provide us with only a single piece of information, its one value. If we consider the different ways in which signals can vary with time, we find that they fall into two distinct categories: those that can change continuously with time, and those that can change only at discrete instants of time. The first kind of signal is called an *analog signal,* while the second is called a *digital signal.* Sometimes, these signals are termed continuous-time and discrete-time, respectively.

As an example of an analog signal, consider a portion of the sound pressure wave that is recorded by a microphone during a musical passage. Figure 2.1.1 shows a sketch of such a signal.

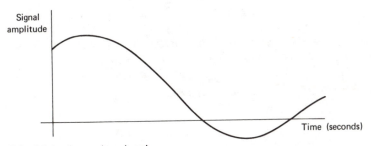

FIG. 2.1.1   An analog signal.

This signal is capable of changing continuously as a function of the abscissa, time, and so conforms to our definition of an analog signal given above. The time axis in this case is simply time, let us say, in seconds.

Now suppose we wish to operate on this signal in some way by using a digital computer. We must put the signal in a different form, since the computer cannot store such a continuously changing function. One very common way to do this is to record sample values of this signal at equally spaced instants. If we sample the signal every 50 microseconds, for example, we obtain the digital signal sketched in Figure 2.1.2.

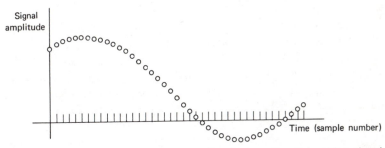

FIG. 2.1.2   The digital signal corresponding to sampling the analog signal in the previous figure.

This sampling interval corresponds to a sampling rate of 20,000 samples/ second, and is a practical rate to use with sound signals. The choice of the sampling rate is an important one, since it determines how accurately the digital signal can represent the original analog signal. In a practical situation, the sampling rate is determined by the range of frequencies present in the original signal, as we shall see in Section 2.4. The operation of obtaining a digital from an analog signal by sampling is usually referred to as analog-to-digital, or A/D, conversion.

One must not get the idea, however, that every digital signal is obtained by sampling an analog signal. As one example of a digital signal not obtained by sampling, consider a rotating radar antenna that receivss a return pulse from an object once every rotation (Fig.2.1.3):

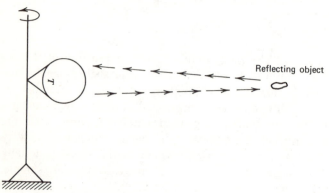

FIG. 2.1.3 A rotating radar antenna.

The strength of the return pulse can be measured only every T seconds, where T is the period of rotation. As a second example, consider the total gain in yardage of a football team per game. Clearly, this quantity can be obtained only once per game, and is a digital signal that bears no relation to any analog signal. A third example arises when we generate a digital signal within a computer for possible conversion to an analog signal. If and where computers learn to talk, the sound waves will probably be generated in this way. The process of producing an analog signal from a digital signal is called, naturally, digital-to-analog, or D/A, conversion and is, in some sense, the reverse of the sampling process discussed above. By tradition, we shall use the term "sample" to refer to a particular valu of a digital signal, even if it was not obtained by the sampling process.

Notice that the time axis of the digital signal shown in Fig. 2.1.2 is labeled simply "sample number." This can be converted to physical time if we know the physical time corresponding to the sampling interval. Often, however, as pointed out above, there is no physical time to which we are restricted. For this reason, we shall always use the term "sample number" to label the successive values of a digital signal. This will avoid our being tied down to any particular time scale in our discussion of digital filters. In some applications, one sampling interval may represent several seconds; in another a nanosecond.

2.2 THE REPRESENTATION OF DIGITAL SIGNALS IN A COMPUTER

A digital signal can be stored in a digital computer in a perfectly natural way, by establishing a one-dimensional array S say, and storing in S(I) the amplitude of sample I. (Sometimes one-dimensional arrays are called "vectors" or "lists.") Such arrays have a first and last element, the first usually being associated with sample number 1, and the last with some array length that is known to us. Thus, every sample value is immediately available to us during a computation by indexing.

Analog signals, on the other hand, have no such natural representation, and are usually sampled to obtain a digital signal if it is necessary to store them in a digital computer. For this reason we shall concentrate on digital signals from now on.

There remains one problem with the storage of digital signals; the accuracy with which their amplitude values can be represented in a digital computer. A sample value is usually stored in a computer as a *word* in the computer's memory. By a word we mean a fixed number of bits that is addressed as a unit by computer operations. Thus, a variable such as S(3) in a FORTRAN statement refers to one word, a set of bits stored at a known location. Let us assume for the purpose of discussion that a word has 32 binary digits. This is representative of the size of words, but some computers may have decimal or octal digits, and may have more or less room in a word. Thus, our word consists of a string of 32 binary storage devices, each of which is *on* or *off*, as in Fig. 1.1.1. As usual, we shall represent 1 by the *on* state, and 0 by the *off* state.

An exception to this occurs when we represent the sign of a number; in this case, we represent " − " by the *on* state, and " + " by the *off* state. Now if we use the first bit as a sign, and the remaining 31 bits as a number in binary form, we can represent all the integers from $-(2^{31}-1)$ to $+(2^{31}-1)$, including zero. These are altogether $2^{32}-1$ numbers; 0 having two representations, $+0$ and -0. For example, the smallest integer is

$$\overset{\displaystyle\overset{\text{Sign}}{\overbrace{}}\quad\overset{\text{31 bits}}{\overbrace{}}}{-11111111111111111111111111111111} \tag{2.2.1}$$

in binary form, or

$$-2{,}147{,}483{,}647 \tag{2.2.2}$$

in decimal form. The largest integer is

$$\overset{\displaystyle\overset{\text{Sign}}{\overbrace{}}\quad\overset{\text{31 bits}}{\overbrace{}}}{+11111111111111111111111111111111} \tag{2.2.3}$$

or

$$2{,}147{,}483{,}647 \tag{2.2.4}$$

in decimal.

Of course, we need not use these $2^{32}-1$ numbers to represent integers; they can represent any set of $2^{32}-1$ numbers that are equally spaced, simply by imagining a decimal point at some fixed location. (Perhaps we should call it here a binary point.) Suppose, for example, that we place an imaginary decimal point before 11 bits. The largest number in this case is

$$\overset{\displaystyle\overset{\text{Sign}}{\overbrace{}}\quad\overset{\text{20 bits}}{\overbrace{}}\quad\overset{\text{11 bits}}{\overbrace{}}}{+11111111111111111111.11111111111} \tag{2.2.5}$$

This number is the number in (2.2.3) divided by 2^{11}. The important consideration is that our entire range of possible numbers is represented by $2^{32}-1$ distinct points, and that these points are *equally spaced* on the real number line. We call this representation *fixed-point*.

The fixed-point representation of real numbers has a very serious drawback, one that sometimes precludes its use in computations with digital signals. That drawback is the need to stay within the range specified by our choice of the decimal point. Suppose, for example, that we choose the decimal point location before 11 bits, as in (2.2.5). The largest number representable is therefore $(2^{31}-1)2^{-11} = 2^{20}-2^{-11}$, which is slightly less than 1,048,576. The distance between successive numbers is constant at 2^{-11}. We say in this case that the *range* is from $-1{,}048{,}576$ to $+1{,}048{,}576$, and the *resolution* is 2^{-11}. If in any computation we multiply two large numbers and a product results that is outside this range, some kind of overflow will take place, and our result will not be correct. This constant need to scale computations so that intermediate results stay in range often discourages the use of fixed-point arithmetic for digital signals.

An alternative to fixed-point representation that largely overcomes this difficulty is the *floating-point* representation. This form corresponds to scientific notation, the first part of the word being used to store the fraction, and the rest to store the exponent. Suppose we reserve the first 24 bits for the fraction and the last 8 for the exponent as a power of 2. We shall allow positive and negative fractions and exponents so that a typical floating-point number would look like

$$\text{Fraction: 24 bits} \qquad \text{Exponent: 8 bits}$$

$$\overbrace{-.110000000000000000000000}\overbrace{+0000011} \qquad (2.2.6)$$
$$\quad \text{Sign} \qquad\qquad\qquad\qquad \text{Sign}$$

Furthermore, we imagine a decimal point at the beginning of the fraction, and always multiply or divide by a power of 2 until a "1" appears in the first position (unless the number is exactly 0). This form is called *normalized*. Thus the number above is in decimal form

$$-(2^{-1}+2^{-2})\times 2^3 = -(4+2) = -6 \qquad (2.2.7)$$

We now consider the question of range. What is the largest number that can be represented in this way? The answer is given by the word

$$\text{24 bits} \qquad\qquad\qquad \text{8 bits}$$

$$\overbrace{+.111111111111111111111111}\overbrace{+1111111} \qquad (2.2.8)$$
$$\quad \text{Sign} \qquad\qquad\qquad\qquad \text{Sign}$$

which is in decimal form

$$(1-2^{-23})\times 2^{(2^7-1)} \approx 2^{127} \qquad (2.2.9)$$
$$\approx 10^{38}$$

which is enormous compared with the largest number representable in the fixed-point representation. The smallest magnitude that can be represented in this normalized floating-point system is in binary form

$$\text{24 bits} \qquad\qquad\qquad \text{8 bits}$$

$$\overbrace{+.100000000000000000000000}\overbrace{-1111111} \qquad (2.2.10)$$

which is in decimal form

$$.5\times 2^{-127} \approx .5\times 10^{-38} \qquad (2.2.11)$$

Thus, the ratio of the largest to the smallest magnitude representable is more than 10^{76}. Of course, each of these positive numbers has a negative counterpart. This range of magnitude makes it possible to do arithmetic operations with real numbers without undo concern about overflow (or underflow), although there are situations in which this becomes a consideration.

It is important to realize that the floating-point representation does not allow for more numbers to be represented than the 2^{32} states of the word. The numbers are, however, spread out nonuniformly, in contrast with the situation in the fixed-point case. For example, when the exponent is 0, the smallest change possible in a floating-point number is 2^{-23}, corresponding to a change in the least significant bit of the fraction. When the exponent is 23, on the other hand, the smallest change possible is 1. Thus, as the floating-point numbers get larger, they get more spread out. Rather than being uniformly distributed on the line, they are spaced so that the resolution is proportional to the absolute magnitude. This makes a great deal of sense in numerical computation, where it is often *relative* size that matters.

To conclude this section, we mention that when more accuracy is required than can be obtained by the single-word floating-point representation, it is possible in many programming languages to define double precision variables, which take two words to store a single real number. We encountered the idea of using two words to store one number in Exercise 1.3.6, where complex variables were used. In this case one word is used to store the real part and one the imaginary part. A double precision complex variable therefore requires four words for its storage.

Example

The fixed-point representation discussed above is called the *sign-magnitude* representation. Other representations are possible, such as the *2's-complement* representation. In this scheme, the positive numbers are the same as in the sign-magnitude scheme. The binary word that would ordinarily come after the largest positive number, however, represents the largest negative number, and successive binary words count up to -1. To illustrate, all 3-bit words and their corresponding numbers are shown below:

Word	*Number*
000	0
001	1
010	2
011	3
100	-4
101	-3
110	-2
111	-1

The following properties of this representation can be checked:

(1) Negative numbers have a 1 in the left-most bit.

(2) A number can be negated by complementing each bit and adding 1.

Another important property concerns the implementation of addition. Suppose we wish to calculate $(3 + 3) + (-4)$ in the 3-bit system above. Treating the entire word as a binary number, and ignoring overflow, we obtain

$$
\begin{array}{r}
011 \\
+011 \\
\hline
110 \\
+100 \\
\hline
010 = 2
\end{array}
$$

which is the correct answer. It is easy to see why this works if we visualize the numbers as being equally spaced on a circle:

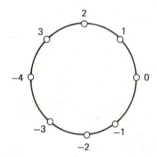

Addition now corresponds to clockwise or counterclockwise motion through the appropriate angle, and the answer will always be correct modulo 8 in the range -4 to $+3$. If the correct answer lies in this range, we will always obtain it by ignoring overflow and treating the entire word as a binary number, as above.

The 2's-complement representation is very convenient for simple implementation of addition, and is used on many minicomputers.

Another fixed-point representation, one that is often used to represent the exponent of floating-point numbers, is the *excess* 2^{n-1} representation for n-bit numbers. In this scheme 2^{n-1} is added to the actual number to be represented and the result is written as a binary integer. For example, consider the 3-bit case, as above. All the 3-bit words and the numbers they represent are shown below:

Word	Number
000	−4
001	−3
010	−2
011	−1
100	0
101	1
110	2
111	3

Notice that we must subtract $2^{n-1} = 4$ from the binary integer to obtain the number it represents, and that the same set of numbers is represented as in the 2's-complement representation. However, this representation does not share with the 2's-complement form the property of giving correct answers for addition modulo 2^n in the range (-2^{n-1}) to $(2^{n-1} - 1)$ if overflow is ignored.

Exercise 2.2.1

Consider the situation where a computer has 6-bit words, and the floating-point representation uses 3 bits for the fraction and 3 for the exponent. Plot every possible positive floating-point number on a linear scale. Compare these with the possible positive fixed-point numbers using the entire 6-bit word. Assume in your solution that only *normalized* floating-point numbers are allowed; that is, that a 1 appears in the first place of the fraction, as assumed in the text.

Exercise 2.2.2

Find out exactly what the bit representations are for fixed-point and floating-point numbers on the computer that you use. How is floating-point zero defined? Discuss the relative advantages and disadvantages of the fixed- and floating-point number representations.

Exercise 2.2.3

Because of the normalization assumed in the text, there are some 32-bit patterns that do not represent floating-point numbers. Hence, there are fewer than 2^{32} distinct floating-point numbers representable by this 32-bit word. Exactly how many distinct floating-point numbers are representable?

Exercise 2.2.4

The floating-point representation used on the IBM 360/91 computer uses hexadecimal notation. That is, numbers are represented to the base 16, with the following correspondence to decimal numbers:

Decimal	Hexadecimal		Decimal	Hexadecimal
0	0		8	8
1	1		9	9
2	2		10	A
3	3		11	B
4	4		12	C
5	5		13	D
6	6		14	E
7	7		15	F

The words are 32 bits long, with each group of 4 bits represented by one hexadecimal digit. The first 2 hexadecimal digits represent the sign of the number and the exponent as follows: if the first two hexadecimal digits are written in binary form, the first bit represents the sign of the number, and the next 7 bits represent the exponent in *excess* $2^6 = 64$ notation. The next 24 bits (or 6 hexadecimal digits) represent the fraction, normalized so that the leading hexadecimal digit is not zero. Translate the following floating point numbers in this representation to decimal:

(a) 41100000

(b) C1111111

(c) 3F200000

(d) 00100000

(e) 7F100000

(f) 80000000

Exercise 2.2.5

Find out how to read in and print out numbers in binary form on your computer, for both fixed- and floating-point.

Exercise 2.2.6

Does *n*-bit binary multiplication, ignoring overflow, give correct results modulo 2^n for the 2's-complement representation? Check your conclusion by working the following multiplication examples in 2's-complement arithmetic, assuming $n = 3$.

    (a) $(-1)(2)$ (d) $(-4)(2)$

    (b) $(1)(-2)$ (e) $(-4)(-3)$

    (c) $(2)(3)$ (f) $(-1)(-2)$

Repeat for excess 4 notation.

Exercise 2.2.7

A 12-bit A/D converter with output in 2's-complement form malfunctions by having bit 8 always *on*, regardless of its true value. (This really happened.) A waveform is sampled, converted, and then reconverted to analog form again using a D/A converter. Describe the result.

Exercise 2.2.8

Let the n bits of a fixed-point number in the sign-magnitude representation be labeled b_0 through b_{n-1}, from right to left. What effect does the following operation have on the number (the left arrow is interpreted as "becomes" or "is set equal to"):

$$b_0 \leftarrow b_1$$
$$b_1 \leftarrow b_2$$
$$\vdots$$
$$b_{n-3} \leftarrow b_{n-2}$$
$$b_{n-2} \leftarrow 0$$
$$b_{n-1} \leftarrow b_{n-1}$$

(This is a right-shift operation, retaining sign.)

Exercise 2.2.9

Answer the same question for the following operation:

$$b_{n-1} \leftarrow b_{n-1}$$
$$b_{n-2} \leftarrow b_{n-3}$$
$$\vdots$$
$$b_1 \leftarrow b_0$$
$$b_0 \leftarrow 0$$

(This is a left-shift operation, retaining sign.)

Exercise 2.2.10

Does a left-shift or a right-shift operation have the same interpretation for the 2's-complement representation?

2.3 QUANTIZING NOISE

Most A/D converters, by nature of their construction, produce fixed-point samples of the input signals. Practical A/D converters that are commercially available usually have from 8 to 14 bits in the fixed-point output, so that the maximum resolution available is about 1 part in $2^{14} = 16,384$. The process of representing the value of the analog signal by a fixed-point number is called *quantization*, and the difference between the actual value of the signal and the fixed-point representation is called *quantization error*, or *quantization noise*.

Let us take up a concrete example of an A/D converter with 12 bits, and assume that the decimal (bit) point is before 5 bits. Thus a fixed-point word looks like

$$7 \text{ bits} \quad 5 \text{ bits}$$
$$\overbrace{\hspace{2cm}} \quad \overbrace{\hspace{1.5cm}}$$
$$-110101.11000$$

$$(2.3.1)$$

which represents the decimal number -53.75. The largest magnitude representable is $(2^{11} - 1)2^{-5} = 64 - 2^{-5}$, and we shall assume that this represents the physical units of volts. Thus, our converter has a range of ± 64 volts, and a resolution of 0.03125 volts. The usual practice in the process of quantization is to round off the actual value of the input analog signal to the nearest number that has a fixed-point representation. Thus, the maximum quantization error that can occur when the signal is within the range of the converter is ½ the resolution. Figure 2.3.1 illustrates the quantization of an analog signal before sampling has taken place.

We shall denote the resolution of an A/D converter by Q. The size of the quantization noise depends quite directly, of course, on the value of Q for the particular converter. The smaller we can afford to make Q, the less quantization noise will be introduced by the conversion process.

We shall now investigate in more detail the dependence of the quantiza-

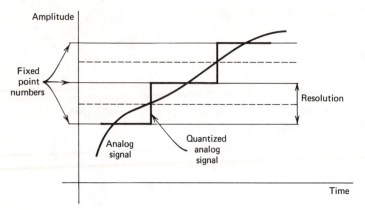

FIG. 2.3.1   Quantization of an analog signal.

tion noise on the resolution Q. For this purpose, we shall consider a typical segment of an analog signal between times when it passes fixed-point numbers. We shall assume that the analog signal is linear in this time, as illustrated in Fig. 2.3.2.

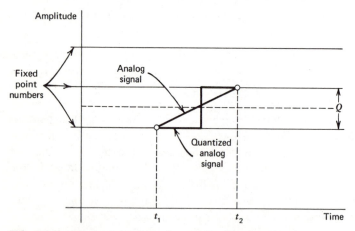

FIG. 2.3.2   A typical segment of a quantized analog signal.

Figure 2.3.3 shows a plot of the quantization error, which goes from zero, to a maximum of $Q/2$, to a minimum value of $-Q/2$, and back to zero.

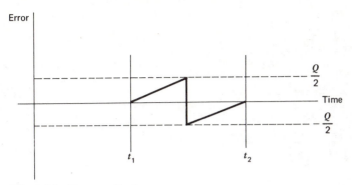

FIG. 2.3.3   The quantization error.

The first question we might ask about this typical segment of the error is: what is its average? Since the error is as often positive as negative, the average value is precisely zero, and this gives us no information about the "size" of the error. A more meaningful measure is the average value of its square, called the *mean-square* value. In this calculation, negative and positive values of the error both count in the positive sense, and the result gives a measure of the size of the squared error we might expect over the segment. The square root of this quantity, called the *root-mean-square* (rms) value, gives us an indication of the size of the error itself. The square of the quantization error is plotted in Fig. 2.3.4.

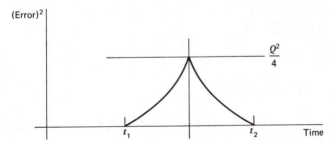

FIG. 2.3.4   The quantization error squared.

We shall now calculate the average value of this waveform. First, we need only consider half the waveform, since both halves have the same shape. Second, we can take the starting time as 0, since this only shifts the waveform without affecting its shape. Hence, we need to find the average value of the waveform shown in Fig. 2.3.5.

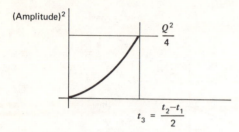

FIG. 2.3.5   Centered version of one-half the error waveform squared.

The function has the mathematical form

$$\left(\frac{t}{t_3}\frac{Q}{2}\right)^2 \tag{2.3.2}$$

where t_3 is half the duration of our original signal segment. To find the average value of this waveform, we calculate

$$\text{Mean-square value} = \frac{1}{t_3}\int_0^{t_3}\left(\frac{t}{t_3}\frac{Q}{2}\right)^2 dt = \frac{Q^2}{4t_3^3}\int_0^{t_3} t^2\,dt = \frac{Q^2}{12} \tag{2.3.3}$$

Notice that the result depends only on Q and not on the length of the segment. The root-mean-square value of the quantizing noise is therefore $Q/\sqrt{12} = 0.29Q$. Thus, if we take our straight-line segment as typical of a segment of the input analog signal, then we may expect a quantizing noise with rms value given by this multiple of the resolution.

To be more precise about the notion of "average" in a statistical sense, we really need to define the concepts of "random signal" and "expectation." The somewhat heuristic argument above will suffice for our present needs. It turns out, however, that the result

$$\text{rms quantization noise} = \frac{Q}{\sqrt{12}} \tag{2.3.4}$$

can be justified for a wide class of random signals as well as our simple straight line segment.

Exercise 2.3.1

Calculate the rms value of the following signals over the intervals indicated:

(a) A sine wave of amplitude 1 over 1 period.

(b) The square wave below, for the interval from 0 to 3.

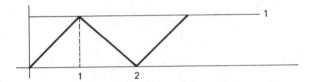

(c) The triangle wave below, for the interval from 0 to 2.

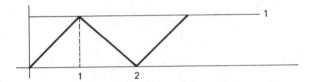

What are the ratios of peak-to-rms for these three cases? What is the significance of this ratio?

Exercise 2.3.2 (computer experiment)

The purpose of this experiment is to measure the rms value of the quantizing noise when a sine wave is sampled, and to compare the measured value to the theoretical value of $Q/\sqrt{12}$.

(1) We shall assume that a hypothetical A/D converter has 4 decimal digits in its fixed-point output, and the decimal point is at the right end so that its range is $-999.$ to $+999.$, and its resolution is 1. Further, we shall assume that the floating-point representation in our computer is accurate enough so that a sine wave generated by the sine function in FORTRAN can be considered to have infinite precision. Generate in a DO-loop the sampled sine wave

$$1000.* \text{SIN (FLOAT (K)} *W)$$

where K is the index parameter of the DO-loop, and W is a parameter of your choice that determines the frequency of the sine wave. W should be small enough so that the argument of the sine wave does not change more than 1/16 of a period between sample values. Also, W should not be an integral fraction of π. Why?

(2) In the DO-loop round the samples of the sine wave off to the nearest integer. This can be done easily by making use of the INT function. Be sure, however, that you *round* to the *nearest* integer, and not *truncate*.

(3) In the DO-loop calculate the value of the quantization error, and its square.

(4) Calculate the average, the average square, and the square-root of the average square quantization error, for 100, 500, and 1000 samples. Compare these with the theoretical values.

If you have time, repeat this experiment for some functions beside the sine wave.

Exercise 2.3.3

Using the methods of this section, calculate the average value of the absolute error. How does this compare with the rms error?

2.4 PHASORS

Having discussed the sampling and quantization processes in some detail, we shall now discuss a particularly important class of digital signals: the sinusoids. The kth sample value of a sinusoid can be written as

$$F(k) = A \cos(\omega k + \phi) \qquad k \text{ is an integer}$$

Figure 2.4.1 shows a plot for nonnegative k.

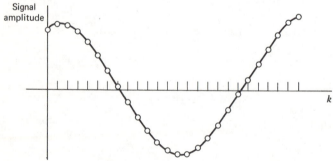

FIG. 2.4.1 A digital sinusoid.

Three parameters characterize the signal: A, ω, and ϕ. A determines the amplitude of the sinusoid, ω the frequency, and ϕ the starting point of the cycle in relation to the origin, $k = 0$. The angle ϕ will be referred to as the *phase* of the sinusoid. We shall now look more closely at the range of values that ω can take on. For this purpose it is more convenient to write $F(k)$ as the real part of a complex exponential, using Euler's formula:

$$F(k) = \text{Real } \{Ae^{j(\omega k + \phi)}\}$$

The function $Ae^{j(\omega k + \phi)}$ is called a *phasor*, and can be interpreted as a point that moves around a circle of radius A in the complex plane. The signal $F(k)$ is the real part of the phasor and, hence, can be interpreted as the projection of the moving point on the real axis. When ω is very small, the point moves slowly (Fig. 2.4.2).

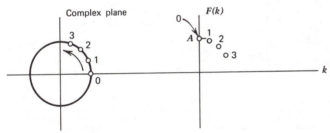

FIG. 2.4.2 A complex digital phasor, together with its real part, a digital sinusoid.

We have arbitrarily taken $\phi = 0$, so that the phasor starts on the real line when $k = 0$. When ω is zero, the phasor remains at its starting point, and the signal is a constant (Fig. 2.4.3).

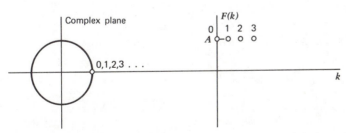

FIG. 2.4.3 A complex digital phasor of frequency 0 radians/sample interval.

If ω is made larger, the point moves around the circle faster. When $\omega = \pi$ the point moves directly from $+A$ to $-A$ and back (Fig. 2.4.4).

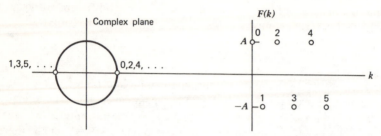

FIG. 2.4.4 A complex digital phasor of frequency π radians/sample interval.

Now suppose that ω is increased beyond π, to $\pi + x$, where x is a small number compared with π. The point then traces the pattern shown in Fig. 2.4.5.

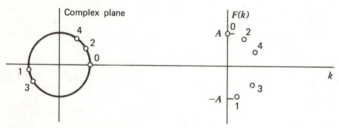

FIG. 2.4.5 A complex digital phasor of frequency $\pi + x$ radians/sample interval, where x is small compared with π.

The point can be thought of as moving a little more than π each sample interval in the counterclockwise direction. Alternatively, the point can be thought of as moving a little *less* than π each sample interval in the clockwise direction. Mathematically, this observation is the result of the following equality:

$$e^{j\omega k} = e^{j(\pi+x)k} = e^{j(2\pi-\pi+x)k} = e^{j(-\pi+x)k}$$

Since $e^{j2\pi k} = 1$, we can subtract or add 2π to the frequency without changing the phasor. Hence a frequency just above π results in the same phasor as a frequency just above $-\pi$. This stroboscopic effect is noticeable in motion picture films of rotating objects, such as locomotive wheels.

A negative frequency means that the phasor is rotating in the clockwise rather than the counterclockwise direction. The real component and, hence, our signal is unchanged. On the other hand, we might have used the imaginary part of the phasor to represent our signal, in which case a negative frequency would result in a difference in sign.

Suppose now that ω is increased still further, until it is slightly less than 2π. The resulting phasor is shown in Fig. 2.4.6.

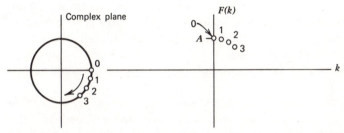

FIG. 2.4.6   A complex digital phasor of frequency slightly less than 2π radians/ sample interval.

We see that the observed frequency has become slightly less than 0. Finally, when $\omega = 2\pi$ the observed frequency is 0. Thus, the observed frequency will always be between $-\pi$ and π, and is obtained by adding or subtracting multiples of 2π to ω until a number in that range is obtained. *The highest frequency that can be represented by a digital signal is therefore π radians/ sample interval.*

Now let us calculate what frequency ω corresponds to if the sampling interval is made to correspond to T seconds of physical time. The period of the phasor $e^{j\omega k}$ can be obtained by setting $\omega K = 2\pi$; K being the number of sample intervals required for a complete cycle of the sinusoid. Hence the period is

$$K = \frac{2\pi}{\omega} \text{ sample intervals/cycle} \qquad (2.4.1)$$

Since each sample interval corresponds to T seconds, the period is

$$\text{Physical period} = KT = \frac{2\pi T}{\omega} \text{ seconds/cycle} \qquad (2.4.2)$$

The actual frequency is the reciprocal of the period, or

$$\text{Physical frequency} = \frac{\omega}{2\pi T} \text{ cycles/second} \qquad (2.4.3)$$

Since $|\omega| \le \pi$, *the highest physical frequency that can be represented by a digital signal is* $1/2T$ cycles/second, where T is the physical time corresponding to a sample interval. This is a fundamental aspect of digital signals. The frequency $1/2T$ cycles/second, or π radians/sample interval, is called the *Nyquist frequency*, after Harry Nyquist who studied telegraph transmission in the 1920's. Notice that $1/T$ cycles/second is the sampling rate, so that the Nyquist frequency is one-half the sampling rate.

A digital phasor can be thought of as arising from the process of sampling at equally spaced times a continuously varying phasor, represented by $e^{j\omega t}$, where t is a continuous variable. Sampling the continuous-time phasor $e^{j\omega t}$, then, means finding its values at times $t = kT$, where k is an integer and T is the interval between sample values (sometimes assumed to be 1).

The following analogy illustrates the sampling process: picture a white disk spinning at a certain speed in a dark room, and a stroboscope that flashes periodically at a certain rate. Assume also that there is a black dot painted near the edge of the disk. The moving dot represents a continuous-time phasor of fixed frequency, while the sequence of illuminated dot positions represents the samples that are values of a discrete-time, digital phasor (Fig. 2.4.7).

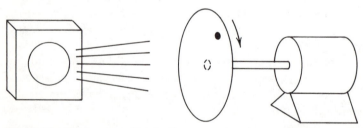

FIG. 2.4.7   Stroboscope illuminating a moving dot: an analogy to the process of sampling a phasor.

If the dot moves slightly more than one-half a revolution between flashes, it will appear to be moving backward at a rate slightly less than one-half a revolution per flash; and in fact the dot will never appear to be moving more than π radians per flash. It then follows that in order to represent accurately the speed of revolution, we must flash (sample) at least twice per period of revolution. If we assume that any analog signal can be thought of as a sum of phasor components, this translates into the following principle (which is another way of looking at the fact that the highest frequency representable by a digital signal is one-half the sampling frequency):

Sampling Principle

In order to represent accurately the frequency components present in an analog signal, the sampling frequency must be at least twice the highest frequency present in the signal.

Example Problem

An analog sine wave generator is connected to an A/D converter whose sampling rate is 10,000 samples/second. This is then connected to a D/A converter, and then to a loudspeaker, as illustrated below:

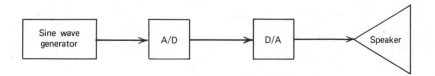

The frequency knob on the sine wave generator is turned slowly from 0 cycles/second to 50,000 cycles/second. Describe the frequency of the resulting sound. Do this by plotting a graph showing the input signal frequency on the horizontal axis, and the frequency of the resulting sound on the vertical axis.

Solution. From the discussion in Section 2.4 we should add or subtract 2π radians/sample interval from the frequency in radians/sample interval until we get a frequency between $-\pi$ and π. This corresponds to adding or subtracting $1/T$ cycles/second, the sampling frequency, from the frequency in cycles/second until we get a frequency between $-1/2T$ and $1/2T$. Thus, the frequency of the resulting sound increases to 5000 cycles/second, jumps to -5000 cycles/second, increases to 5000 cycles/second, and so on, as shown in Fig. 2.4.8. Of course the ear perceives no difference between a positive and a negative frequency, so we plot just the magnitude of the frequency.

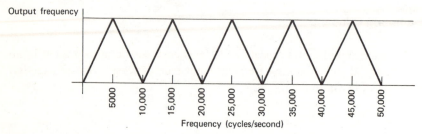

FIG. 2.4.8 Output frequency *vs.* input frequency in the example.

The process whereby an analog sine wave of a certain frequency is sampled and interpreted as a sine wave of lower frequency is called *aliasing,* presumably because the digital signal could have been obtained from many different original analog signals, all of which can be considered aliases.

Because of the aliasing effect it is usually necessary to precede A/D conversion by a continuous-time filter that removes all frequencies above the Nyquist frequency. Such a "low pass" continuous-time filter is called a *prefilter*; the theory of the operation of such a device must be postponed until we study electrical networks in Chapter 11.

In a similar way, the process of D/A conversion introduces unwanted frequencies *above* the Nyquist frequency, and it is necessary to follow D/A conversion by another low pass operation to remove all frequencies above the Nyquist frequency. This filter is frequently called a *postfilter*. Thus, the system drawn above will in practice include two continuous-time filters, as shown below:

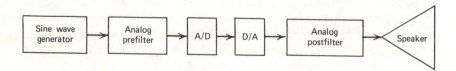

Exercise 2.4.1

An analog musical passage contains frequencies up to 10,000 cycles/second, and is to be sampled for conversion to a digital signal. What range of sampling frequencies will allow all the frequencies in the original signal to be represented in the digital signal? What would happen if too low a sampling frequency were used?

Exercise 2.4.2

Let us assume that an analog signal representing music can be satisfactorily quantized with a 12 bit A/D converter, and that it contains frequencies only up to 10,000 cycles/second. If a Beethoven symphony is sampled at a rate determined in the previous exercise, using this converter, how many bits are needed to store the result? Make a reasonable assumption about the length of the symphony.

Exercise 2.4.3

(a) We wish to buy an A/D converter for converting seismographic signals to digital form so that we can use a computer to analyze them. The pressure recording instrument has an amplitude range of $\pm$ 10 volts. The background noise has an rms value of about 0.05 volts. We have a choice of buying converters with 7, 12, or 16 significant bits, at costs of $500, $1000, or $2000, respectively. Which is the least expensive converter that is adequate for our use? Explain your reasoning.

(b) The highest frequency in a seismic signal that we are interested in is 20 Hz. Choose a sampling rate and explain your choice.

(c) We now want to send our seismic signal in digitized form over a data link with a capacity of 1000 bits/second. Is one such channel adequate?

Exercise 2.4.4

Color video TV signals can presently be sampled and quantized using 8 bits/sample and a 12 MegaHertz sampling rate (12×10^6 samples/second). The signal can be considered to have frequencies from 0 to 5.5 MegaHertz.

(a) Discuss this choice of sampling rate in terms of the Nyquist frequency.

(b) Estimate the rms error in the signal caused by quantizing, assuming the signal to vary between $-1/2$ and $+1/2$ volt.

Exercise 2.4.5

A digital communication link carries binary coded words representing samples of an input signal X. The link is operated at 10,000 bits/second, and each input sample is quantized into 1024 different voltage levels. What is

(a) The sampling frequency?

(b) The Nyquist frequency?

(c) The highest frequency of the analog input signal that can be represented in the digital signal?

(d) The frequency in radians/sample interval corresponding to 100 Hz?

(e) The root-mean-square error caused by quantizing noise if the voltage range of the input is 0 to 10.24 volts?

Exercise 2.4.6

The term Nyquist frequency is sometimes used to mean the highest frequency present in an analog signal to be sampled; and the term *folding frequency* is used to mean one-half the actual sampling frequency. Explain the choice of the term "folding" frequency.

Further Reading

An early discussion of quantizing, coding, and implementation of the sampling process can be found in the reference below.

1. *Notes on Analog-Digital Conversion,* A. K. Susskind (editor), MIT Press, Cambridge, Mass., 1957.

The reader will find a more recent treatment in Chapter 1 of the next citation.

2. *Discrete-Time and Computer Control Systems,* J. A. Cadzrow and H. R. Martens, Prentice-Hall, Englewood Cliffs, N.J., 1970.

Also see Chapter 7, "Communications and Data Conversion Circuits," in citation 3:

3. *Digital Electronics with Engineering Applications,* T. P. Sifferlen and V. Vartanian, Prentice-Hall, Englewood Cliffs, N.J., 1970.

For a complete discussion of number representations, see citation 4:

4. *The Art of Computer Programming;* vol II: *Seminumerical Algorithms,* D. E. Knuth, Addison-Wesley, Reading, Mass., 1969.

3.
MOVING AVERAGE DIGITAL FILTERS

3.1 A SIMPLE FILTER

There are many situations in which we wish to operate on a given digital signal to produce another signal. For example, we may wish to filter out some noise (to the extent it is possible); or we may wish to extract signal components within a certain range of frequencies; or may wish to predict future values. Such processing is generally termed *digital filtering*, and the algorithm that effects the processing is termed a *digital filter*. We begin by considering an especially simple filter whose purpose it is to smooth short term irregularities in a signal.

Suppose a digital signal has irregular fluctuations of a random nature that are superimposed on an otherwise smooth waveform, as shown in Fig. 3.1.1.

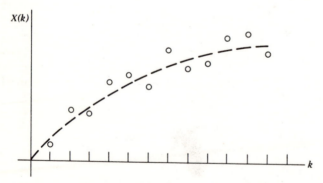

FIG. 3.1.1   A digital signal with irregular fluctuations.

One scheme to smooth such a signal is to average the adjacent sample values. Since the underlying waveform (the signal) is assumed to be smooth, this averaging process will tend to reduce the amplitude of the irregular component (the noise). If $X(I)$ is the original waveform, and $Y(I)$ is the smoothed waveform, the averaging process can be expressed in FORTRAN as

$$\text{DO } 10 \;\; I=2,N$$
$$10 \;\; Y(I) = .5 * X(I) + .5 * X(I-1) \tag{3.1.1}$$

where N is the total length of the signal record. That is, the present value of Y is the average of the present value of X and the preceding value of X. Notice that this leaves the first value of Y undefined, since $X(0)$ is not given. The result is sketched in Fig. 3.1.2.

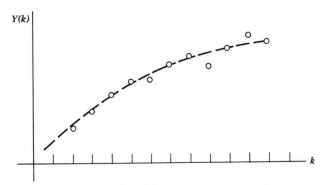

FIG. 3.1.2   A smoothed version of the previous digital signal.

This process is one example of a digital filter; and X is referred to as the *input,* and Y is referred to as the *output.* If we call this filter H, we write $Y = H(X)$, and represent it as in Fig. 3.1.3.

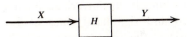

FIG. 3.1.3   Representation of digital filter H with input signal X and output signal Y.

This filter has two very important properties. First, it is linear, which means that if X' and X'' are any two possible input signals, then

$$H(aX' + bX'') = aH(X') + bH(X'') \tag{3.1.2}$$

where a and b are any numbers. This property follows directly from the defining equation (3.1.1), and implies that if the input is broken down into components, the response of the filter to the total input is the sum of the responses to each component. Figure 3.1.4 shows in diagramatic form the meaning of linearity; Fig. 3.1.4a generates as output the left-hand side of Eq. 3.1.2, while Fig. 3.1.4b generates the right-hand side of Eq. 3.1.2. The filter is linear if these outputs are equal for all choices of X', X'', a, and b.

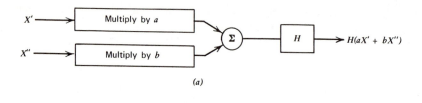

(a)

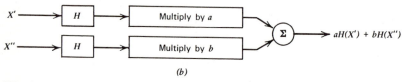

(b)

FIG. 3.1.4   Illustration of linearity: the results of the operations in (a) and (b) are identical for a linear filter H.

The second property is called *time-* or *shift*-invariance, and means that if the input is shifted by a certain number of sample intervals, the output is also. That is:

$$H(X(I+K)) = Y(I+K) \qquad (3.1.3)$$

where $Y = H(X)$, and K is any integer. Again, it follows directly from the defining equation for H, Eq. 3.1.1, that this filter is time-invariant. Figures 3.1.5a and b show, respectively, the left- and right-hand sides of Eq. 3.1.3. For any filter H to be time-invariant, the two outputs must be equal for every integer K.

Linear, time-invariant (LTI) filters have many convenient properties not shared by other types of filters, especially properties having to do with the response to sinusoidal inputs and to the sinusoidal components of more general signals. For this reason, we shall deal almost exclusively with LTI filters from now on.

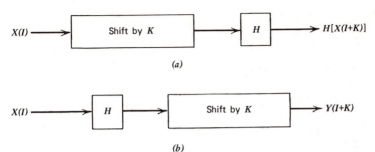

FIG. 3.1.5 Illustration of time-invariance: the results of the operations in (*a*) and (*b*) are identical for a time-invariant filter *H*.

Example Problem

Determine whether the filter H defined by

$$Y(k) = C^k X(k) \qquad \text{all } k, \qquad C \text{ is a constant}$$

is (a) linear; (b) time-invariant.

Solution. (a) The left-hand side of Eq. 3.1.2 is here

$$C^k[aX'(k) + bX''(k)] = aC^k X'(k) + bC^k X''(k)$$

while the right-hand side yields

$$aC^k X'(k) + bC^k X''(k)$$

which is the same. Hence H is linear.

(b) The left-hand side of Eq. 3.1.3 yields

$$C^k X(k + K)$$

whereas the right-hand side yields

$$[C^k X(k)]_{k \text{ replaced by } k+K} = C^{k+K} X(k + K)$$

Therefore H is not time-invariant.

Exercise 3.1.1

Determine whether each of the following filters is (1) linear; (2) time-invariant:

(a) $Y(k) = [X(k)]^2$, all k

(b) $Y(k) = X(k) + X(k-100)$, all k

(c) $Y(k) = \begin{cases} 0, k < 0 \\ X(k) + X(k-100), k \geq 0 \end{cases}$

(d) $Y(k) = X(k) * X(k-1)$, all k

(e) $Y(k) = X(2k)$, all k

(f) $Y(k) = \begin{cases} X(k), \text{ if } X(k) \geq 0 \\ 0, \text{ if } X(k) < 0 \end{cases}$

(g) $Y(k) = kX(k)$, all k

(h) $Y(k) = k + X(k) + X(k-1)$, all k

(i) $Y(k) = X(k+1)$, all k

(j) $Y(k) = \begin{cases} X(k-1), k \text{ even} \\ X(k-3), k \text{ odd} \end{cases}$

(k) $Y(k) = e^{jX(k)}$, all k

(l) $Y(k) = |X(k)|$, all k

3.2 THE FREQUENCY RESPONSE OF OUR SIMPLE FILTER

Consider now the effect of the simple filter described above on a sinusoidal input X. We shall assume that X is a phasor of frequency ω:

$$X(k) = e^{j\omega k} \qquad (3.2.1)$$

and consider only the real part of signals whenever we wish to relate them to physical signals. The output Y is determined by the equation of our filter:

$$Y(k) = .5 * X(k) + .5 * X(k-1) \qquad (3.2.2)$$

Substituting the first equation in the second, we obtain

$$\begin{aligned} Y(k) &= .5 \, e^{j\omega k} + .5 \, e^{j\omega(k-1)} \\ &= .5(1 + e^{-j\omega}) e^{j\omega k} \\ &= .5(1 + e^{-j\omega}) X(k) \end{aligned} \qquad (3.2.3)$$

If we define the function

$$H(\omega) = .5(1 + e^{-j\omega}) \qquad (3.2.4)$$

the above equation takes on the simple form:

$$Y(k) = H(\omega) X(k) \qquad (3.2.5)$$

This equation has the following interpretation: $H(\omega)$ is a complex number, which, in polar form, has magnitude $|H(\omega)|$ and angle Arg $H(\omega)$; both of which depend on the frequency of the input ω. Hence, $Y(k)$ is also a phasor, with the same frequency as the input $X(k)$. The effect of the filter H is therefore to multiply the amplitude of the input sinusoid by $|H(\omega)|$ and to shift its phase by Arg $H(\omega)$. The function $H(\omega)$ is called the *transfer function* of the digital filter H. If an arbitrary input is thought of as consisting of a sum of sinusoidal components, then the effect of the filter is to act on each sinusoidal component in the manner just described.

The effect of a digital filter on a sinusoidal input signal motivates the following terminology. The transfer function $H(\omega)$ will sometimes be referred to as the *frequency response* of the filter H. When considered as a complex function of frequency, we shall call its magnitude the *amplitude, amplitude characteristic,* or *amplitude response;* and its angle the *phase, phase characteristic,* or *phase response* of the filter or transfer function.

Let us now calculate what the amplitude and phase of H are. The amplitude is

$$
\begin{aligned}
|H(\omega)| &= |.5(1+e^{-j\omega})| \\
&= |.5\,e^{-j\omega/2}\,(e^{j\omega/2}+e^{-j\omega/2})| \\
&= |.5||e^{-j\omega/2}|\ |2\cos\frac{\omega}{2}| \\
&= |\cos\frac{\omega}{2}|
\end{aligned}
$$

This function can be plotted as a function of the frequency as shown in Fig. 3.2.1.

We need only plot this function for frequencies in the range from 0 to π, since its values for negative frequencies are the mirror image of those for

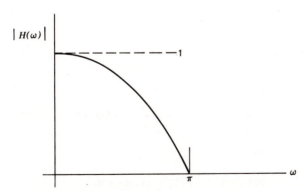

FIG. 3.2.1   The amplitude of the low pass filter H.

positive frequencies, and π is the largest frequency that can be present in a digital signal. Notice that the amplitude of the transfer function is a decreasing function of frequency; which means that the higher the frequency of an input signal, the more it is attenuated. Such a filter is called *low pass*. We can check the values at $\omega = 0$ and $\omega = \pi$ as follows: at $\omega = 0$, the input X is simply the constant 1 at every sample number. Hence Y, which is the average of adjacent values of X, is also 1; so that the amplitude of the filter transfer function is 1, verifying Fig. 3.2.1. At $\omega = \pi$, the input X is alternately 1 and -1, so that Y is always 0, again in agreement with our calculation.

Now let us consider the phase of the filter H:

$$\text{Arg } H(\omega) = \text{Arg}[e^{-j\omega/2} \cos \frac{\omega}{2}]$$

$$= -\frac{\omega}{2}$$

Hence, the phase is simply one-half the frequency. Since ω represents the number of radians per sample interval, the phase represents a delay of one-half a sample interval. This is reasonable, since Y is the average of the present input and the value of the input one sample interval in the past. Figure 3.2.2 shows the phase of H as a function of frequency.

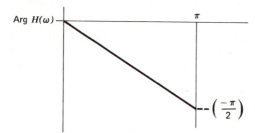

FIG. 3.2.2   The phase of the low pass filter H.

Now let us consider another filter, say G, which is defined by the following input-output relation:

$$Y(k) = .5 * X(k) - .5 * X(k-1)$$

which is the same as H except that the difference instead of the sum is taken. This has the effect of accentuating rather than smoothing random fluctuations in the input signal. The transfer function of G can be calculated in exactly the same manner as above, yielding

$$G(\omega) = .5\,(1 - e^{-j\omega})$$

This time the magnitude is

$$|G(\omega)| = \left|\sin \frac{\omega}{2}\right|$$

and the phase is

$$\text{Arg } G(\omega) = \text{Arg}[e^{-j\omega/2}\, j \sin \frac{\omega}{2}\,]$$

$$= \frac{\pi}{2} - \frac{\omega}{2}$$

These are plotted in Fig. 3.2.3.

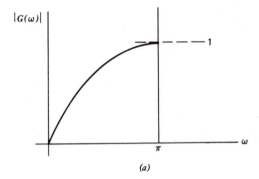

(a)

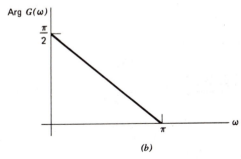

(b)

FIG. 3.2.3   (a) The amplitude and (b) the phase of the high pass filter G.

G is hence a *high pass* filter, since it attenuates lower frequencies more than high ones. At low frequencies the phase of the filter is close to $\pi/2$, which means it tends to produce a sinusoid that occurs one-quarter cycle sooner then the input.

Example

One should not infer from the two examples given thus far that the transfer function can always be simplified by factoring out $e^{-j\omega/2}$. Consider, for example, the moving average filter defined by the equation

$$Y(k) = .6 * X(k) + .4 * X(k-1) \qquad (3.2.6)$$

The transfer function of this filter, obtained after substituting

$$X(k) = e^{jk\omega} \qquad (3.2.7)$$

is

$$H(\omega) = .6 + .4e^{-j\omega} \qquad (3.2.8)$$

Factoring out $e^{-j\omega/2}$, we obtain

$$H(\omega) = e^{-j\omega/2}(.6e^{j\omega/2} + .4e^{-j\omega/2}) \qquad (3.2.9)$$

but we cannot use Euler's formula for the factor in parentheses to obtain a factor equal to $\cos(\omega/2)$ or $\sin(\omega/2)$. We can, however, use Euler's formula directly in (3.2.8) to obtain

$$H(\omega) = .6 + .4 \cos \omega - .4j \sin \omega \qquad (3.2.10)$$

and

$$\text{Real } H(\omega) = .6 + .4 \cos \omega$$
$$\text{Imag } H(\omega) = -.4 \sin \omega \qquad (3.2.11)$$

The magnitude of the transfer function is given by

$$\begin{aligned}
|H(\omega)| \quad &= \sqrt{[\text{Real } H(\omega)]^2 + [\text{Imag } H(\omega)]^2} \\
&= \sqrt{[.6 + .4 \cos \omega]^2 + [.4 \sin \omega]^2} \\
&= \sqrt{.36 + .48 \cos \omega + .16 \cos^2 \omega + .16 \sin^2 \omega} \\
&= \sqrt{.52 + .48 \cos \omega}
\end{aligned} \qquad (3.2.12)$$

and the phase angle is given by

$$\begin{aligned}
\text{Arg } H(\omega) &= \arctan \left[\frac{\text{Imag } H(\omega)}{\text{Real } H(\omega)} \right] \\
&= \arctan \left[\frac{-.4 \sin \omega}{.6 + .4 \cos \omega} \right]
\end{aligned} \qquad (3.2.13)$$

These are sketched in Fig. 3.2.4

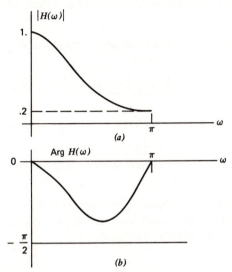

FIG. 3.2.4   (*a*) The amplitude and (*b*) the phase of the filter in the example.

Exercise 3.2.1

Calculate and plot the amplitude and phase of the following filters:

(a) $Y(k) = .25 * (X(k) + 2 * X(k - 1) + X(k - 2))$
(b) $Y(k) = .25 * (X(k) - 2 * X(k-1) + X(k-2))$
(c) $Y(k) = .5 * (X(k) + X(k-2))$
(d) $Y(k) = .5 * (X(k+1) + X(k-1))$
(e) $Y(k) = .125 * (X(k) + 3 * X(k-1) + 3 * X(k-2) + X(k-3))$
(f) $Y(k) = X(k) + X(k-8)$
(g) $Y(k) = X(k) - X(k-8)$

Exercise 3.2.2 (computer experiment)

The purpose of this experiment is to implement the low pass moving average digital filter discussed in Sections 3.1 and 3.2, and to verify its frequency response characteristics by applying sine waves of varying frequencies.

1. Dimension an array $X(200)$, which will hold sample values of our input signal X.

2. Dimension an array $Y(200)$, which will hold sample values of our computed output signal Y.

3. In a DO-loop, set up the input signal X to be a sine wave of frequency W of your choosing.

4. In another DO-loop, calculate Y using the moving average filter of (3.1.1).

5. Print out the input and corresponding output sample values in two parallel columns.

6. The amplitude and phase of X are known. Plot the signal Y, and estimate its amplitude and phase. Check this result with the amplitude and phase response of the low pass filter, as we derived in Section 3.2.

Exercise 3.2.3

A moving average digital filter has the defining equation

$$Y(k) = X(k) - X(k-8)$$

This filter has zero output when phasors of certain frequencies are applied.

(a) What are these frequencies in radians/sample interval?

(b) What are these frequencies in Hz, if the sampling frequency is 10,000 samples/second?

3.3 MOVING AVERAGE FILTERS

The filters described above can be generalized so that each output sample is obtained by averaging M sample values of the input. Such a filter is described by the equation

$$Y(K) = C(1)X(K) + C(2)X(K-1) + \ldots + C(M)X(K-M+1)$$
$$\text{(3.3.1)}$$

where the weighting coefficients are denoted by $C(I)$. This is expressed simply in FORTRAN as

```
DO   10   K=M,N
Y(K)=0.
DO   10   J=1,M
10   Y(K)=Y(K)+C(J)*X(K−J+1)
```

where, again, N is the number of input sample values available. There are now $M-1$ undefined sample values of the output: $Y(1)$ to $Y(M-1)$. The simple averaging filter described in the previous two sections corresponded to the case $M=2$, with $C(1) = C(2) = .5$. Such filters are called *moving average* filters of order $M - 1$. Thus a zero order moving average filter consists of just multiplication by a constant, $C(1)$. The order of such filters corresponds to the number of *past* inputs that affect the present output.

The frequency response of a moving average filter can be calculated in the same way as in the special cases considered before. Letting $X(k) = e^{j\omega k}$ in (3.3.1), we obtain

$$Y(k) = [C(1) + C(2)e^{-j\omega} + \ldots + C(M)e^{-j(M-1)\omega}]X(k)$$

so that the transfer function is

$$H(\omega) = C(1) + C(2)e^{-j\omega} + \ldots + C(M)e^{-j(M-1)\omega} \qquad (3.3.2)$$

where we shall call the moving average filter H. We see now that the transfer function of any moving average filter will be a function of the variable $e^{j\omega}$, and will depend on ω only in this way. Therefore, for convenience, we shall use the symbol z in place of $e^{j\omega}$, and write

$$H(z) = C(1) + C(2)z^{-1} + \ldots + C(M)z^{-(M-1)} \qquad (3.3.3)$$

Notice that this is an abuse of notation, since we should write $H(-j \ln z) = H(\omega)$. There is no danger of confusion, however, since we shall use the letters ω and z only in this way from now on. To summarize: *any moving average digital filter has a transfer function that is a polynomial in non-positive powers of z.*

This brings us to a point that we have ignored: can the present value of Y depend on *future* values of the input X? Insofar as computer instructions are concerned, there is no reason to prefer the past to the future. However, if the filter is envisioned as operating on digital signals as they become available (i.e., in *real time*), then it is impossible for the present output to depend on future inputs. In this case, we say that the filter is *realizable,* that is, it does not use information not available to it. Realizable filters have transfer functions of the form (3.3.3) and do not contain positive powers of z. Hence, our original treatment of moving average filters in (3.3.1) dealt only with the realizable case. All this is not very critical, since we can always delay the output enough so that the required inputs become available. Since $Y(k)$ is a phasor of the form $Ae^{j(\omega k+\phi)}$, a delay of m sample intervals corresponds to multiplication by z^{-m}. Hence, delaying Y so that the filter becomes realizable corresponds to multiplying the transfer function by a power of z sufficiently negative so that all the terms in the transfer function have nonpositive powers of z.

We see from the preceding discussion that z can be interpreted as a kind of *operator* that shifts signals to the left (sooner) upon multiplication. Multiplication by a negative power of z corresponds to a shift to the right (later). This provides a simple way of interpreting the transfer function of a moving average filter: every term of the form az^{-n} means that $Y(k)$ contains the term $aX(k-n)$, which is the input delayed by n units and multiplied by a.

In practice, moving average filters with a great many terms are common; a 251st order filter, for example, is quite practical (see Fig. 3.3.1). Such filters are usually designed to produce a desired frequency response, but we shall not deal here with the problem of how this is done.

Example Problem

The following method of predicting a signal is to be implemented in connection with a radar system for tracking aircraft: given the measured position of an airplane at sampling times $(k-2)$ and $(k-1)$, say $X(k-1)$ and $X(k-2)$, the predicted position at time k is obtained by extending a straight line through $X(k-2)$ and $X(k-1)$ and finding the height of this straight line at time k. Call the predicted position $Y(k)$.

(a) Show that the relation between the digital signals X and Y is that of a linear time-invariant filter.

(b) Find the transfer function of the filter.

(c) Sketch the amplitude of this transfer function as a function of the frequency.

Solution. The fact that the relation between X and Y represents linear time-invariant filtering will follow from the fact that the relation is in fact a

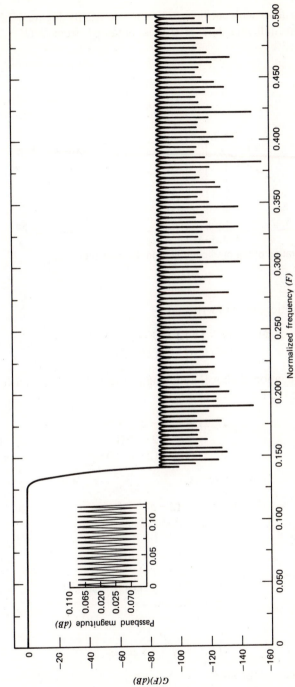

FIG. 3.3.1 The magnitude of the transfer function of a 251st order low pass moving average filter. The frequency $F=0.5$ corresponds to the Nyquist frequency. (From "A New Technique for the Design of Non-Recursive Digital Filters," E. Hofstetter, A. Oppenheim, and J. Siegel, *Proc. 5th Annual Princeton Conference on Information Sciences and Systems*, pp. 64–72; Princeton, N.J., March 1971; with permission of the authors.)

moving average filter with coefficients that are constants. Figure 3.3.2 shows a diagram that indicates how $Y(k)$ is obtained from $X(k-1)$ and $X(k-2)$.

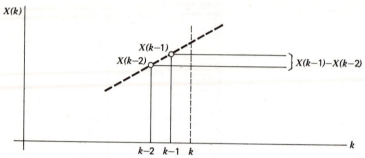

FIG. 3.3.2   Operation of the prediction filter in the example.

From the illustration, we can write

$$Y(k) = X(k-1) + [X(k-1) - X(k-2)]$$
$$= 2*X(k-1) - X(k-2) \qquad (3.3.4)$$

This is, as we mentioned above, a moving average filter. The transfer function is, according to Eqs. 3.3.2 and 3.3.3

$$H(z) = 2z^{-1} - z^{-2}$$
$$= z^{-1}(2 - z^{-1})$$
$$= e^{-j\omega}(2 - e^{-j\omega}) \qquad (3.3.5)$$

The magnitude of $H(\omega)$ squared is

$$|H(\omega)|^2 = |e^{-j\omega}(2 - e^{-j\omega})|^2$$
$$= |e^{-j\omega}|^2 |2 - e^{j\omega}|^2$$
$$= |2 - e^{-j\omega}|^2$$
$$= |2 - \cos\omega + j\sin\omega|^2$$
$$= (2 - \cos\omega)^2 + \sin^2\omega \qquad (3.3.6)$$
$$= 4 - 4\cos\omega + \cos^2\omega + \sin^2\omega$$
$$= 5 - 4\cos\omega$$

Therefore

$$|H(\omega)| = \sqrt{5 - 4\cos\omega} \qquad (3.3.7)$$

which is sketched in Fig. 3.3.3.

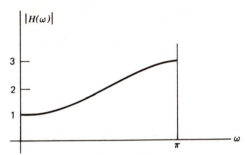

FIG. 3.3.3   The magnitude of the transfer function of the prediction filter.

Exercise 3.3.1

Let the coefficients of a moving average filter be symmetric about their center. That is, let $C(1) = C(M)$, $C(2) = C(M - 1)$, and so on. Show that the resulting filter has a phase that is a decreasing function of ω with a constant slope, except for a finite number of points at which jumps by π, or 2π take place, the jumps by 2π being necessary only to keep the phase in the range $-\pi$ to π.

Exercise 3.3.2

Let H' and H'' be two moving average filters, with transfer functions $H'(z)$ and $H''(z)$, respectively. Consider the filter in Fig. 3.3.4, which is defined by the result of filtering first by H' and then by H''. Prove that the transfer function of this new filter is just $H'(z)H''(z)$.

FIG. 3.3.4   Cascade of two moving average filters, Exercise 3.3.2.

Exercise 3.3.3

Let the filter H' have the transfer function of Eq. 3.3.3, and let H'' have the same coefficients in reverse order:

$$H''(z) = C(M) + C(M-1)z^{-1} + \ldots + C(1)z^{-(M-1)}$$

How are the amplitude functions of $H'(z)$ and $H''(z)$ related? The phases?

Exercise 3.3.4

Consider the digital filter obtained by adding the outputs of n moving average digital filters that have the same input, as shown below:

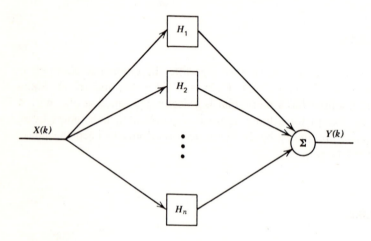

(a) Show that the overall filter is a moving average filter.
(b) Show that the transfer function of the overall filter is

$$\sum_{k=1}^{n} H_k(z)$$

where the individual filters have transfer functions $H_k(z)$, $k = 1, \ldots, n$.

Exercise 3.3.5

Let $H(z)$ and $G(z)$ be the transfer functions of moving average digital filters. Show that $H[1/G(z)]$ is also a moving average digital filter. Is it the same as the filter with transfer function $H(z)G(z)$?

Exercise 3.3.6

Calculate and plot the amplitude and phase of the moving average digital filters with the following transfer functions:

(a) $H(z) = 2 + z^{-2}$
(b) $H(z) = (z^{-1} + z^{-2})/2$
(c) $H(z) = z^{-1}$
(d) $H(z) = (z + z^{-1})/2$

Exercise 3.3.7

Show that a moving average filter is linear and time-invariant.

3.4 A COMPUTER PROGRAM FOR CALCULATING THE FREQUENCY RESPONSE OF MOVING AVERAGE FILTERS

We are now in a position to write a FORTRAN program that will calculate the amplitude and phase of an arbitrary moving average digital filter, with a transfer function given by (3.3.3). Such a program is shown in Fig. 3.4.1. We begin by reading in M, the number of coefficients of the filter, which in this case is one more than the order of the filter. Next, with the statement

$$\text{READ(5,4) (C(I), I=1,M)} \qquad (3.4.1)$$

we read in the coefficients. These are then printed, so that we have a record of the input data.

The DO 7 loop calculates the frequency response at each of 101 points determined by the frequency variable

$$\text{F} = .01 * \text{FLOAT(J} - 1) \qquad (3.4.2)$$

As J goes from 1 to 101, F takes on the values 0. to 1. in steps of .01. F is interpreted as the *frequency in fractions of the Nyquist frequency,* so that the radian frequency is F multiplied by π. The statement

$$\text{ZINV} = \text{CEXP(CMPLX(0., }-1.* \text{F} * \text{PI)}) \qquad (3.4.3)$$

calculates the complex variable

$$z^{-1} = e^{-jF\pi} \qquad (3.4.4)$$

```
C...THIS PROGRAM CALCULATES AMPLITUDE AND PHASE OF A MOVING
C...AVERAGE DIGITAL FILTER WITH TRANSFER FUNCTION
C...C(1)+C(2)*Z**(-1)+...+C(M)*Z**(-1*(M-1))
      COMPLEX ZINV,G,P
      COMPLEX CEXP,CMPLX
      DIMENSION C(24)
      PI=3.14159265
      CONVRT=180./PI
    1 READ(5,2)M
    2 FORMAT(I2)
      WRITE(6,3)M
    3 FORMAT('1MOVING AVERAGE FILTER FREQUENCY RESPONSE, M=',I4)
      READ(5,4)(C(I),I=1,M)
    4 FORMAT(8F10.5)
      WRITE(6,5)(C(I),I=1,M)
    5 FORMAT(' COEFFICIENTS ARE'/8(1X,F10.5))
      DO 7 J=1,101
      F=.01*FLOAT(J-1)
C......F IS FREQUENCY IN FRACTIONS OF THE NYQUIST FREQUENCY
      ZINV=CEXP(CMPLX(0.,-1.*F*PI))
      P=CMPLX(1.,0.)
      G=CMPLX(0.,0.)
      DO 6 I=1,M
      G=G+P*C(I)
    6 P=P*ZINV
      AMP=CABS(G)
      X=REAL(G)
      Y=AIMAG(G)
      IF(ABS(X).GT.1.E-8)GOTO61
      PHASE=SIGN(90.,Y)
      GOTO7
   61 PHASE=ATAN2(Y,X)*CONVRT
    7 WRITE(6,8)F,AMP,PHASE
    8 FORMAT(' F=',F6.3,' AMP=',E16.8,' PHASE=',E16.8)
      GOTO1
      END
```

FIG. 3.4.1   A FORTRAN program for calculating the amplitude and phase of a moving average digital filter.

at the frequency of interest. Next, the DO 6 loop is used to add up the terms

$$C(I) * z^{-I} \tag{3.4.5}$$

yielding the transfer function G. The amplitude and phase are calculated using built-in FORTRAN functions, as follows:

$$AMP = CABS(G) \tag{3.4.6}$$

$$X = REAL(G) \tag{3.4.7}$$

$$Y = AIMAG(G) \tag{3.4.8}$$

$$PHASE = ATAN2(Y,X) * CONVRT \tag{3.4.9}$$

The built-in function ATAN2(Y,X) yields the arctan of the number in the plane with ordinate Y and abscissa X, producing a number between $-\pi$ and π. The factor

$$CONVRT = 180./PI \qquad (3.4.10)$$

defined at the beginning of the program converts this from radians to degrees. The absolute value of X is tested, and if it is less than 10^{-8}, the arctan is set equal to 90° times the sign of Y. This special case may or may not be handled correctly by the built-in ATAN2 function. The final statement in the main DO 7 loop writes out the normalized frequency F, and the amplitude and phase at that frequency. When the main loop is completed, the program transfers control to the beginning again, where more data will be read if it is supplied. Otherwise, an END-OF-FILE will be encountered, and execution will be automatically terminated.

Figure 3.4.2 shows the results obtained when the following moving average filter transfer function was analyzed using this program:

$$.068 - .106z^{-2} + .318z^{-4} + .5z^{-5} + .318z^{-6} - .106z^{-8} + .068z^{-10}$$
$$(3.4.11)$$

The amplitude plot shows that the filter is an approximation to a low pass filter that passes only those frequencies below ½ the Nyquist frequency (F = .5). The coefficients of this filter are symmetrical about the center coefficient, as in Exercise 3.3.1, and the phase plot verifies that the phase is a linearly decreasing function of the frequency, except for jumps of size π or 2π.

3.5 ZEROS

Since the transfer function of a moving average filter is a polynomial in the variable $z^{-1} = e^{-j\omega}$, we can find the roots and write the polynomial as a product of first order factors. Thus, (3.3.1) becomes

$$H(z) = C(1) + C(2)z^{-1} + \ldots + C(M)z^{-(M-1)}$$
$$= C(1)z^{-(M-1)}\left(z^{M-1} + \frac{C(2)}{C(1)}z^{M-2} + \ldots + \frac{C(M)}{C(1)}\right) \qquad (3.5.1)$$
$$= C(1)z^{-(M-1)}(z - z_1)(z - z_2)\ldots(z - z_{M-1})$$

Notice that this is the transfer function of an $(M-1)$st order filter, and that the polynomial that needs to be factored is of order $M-1$. We have factored out the first coefficient $C(1)$, assumed nonzero, and the factor $z^{-(M-1)}$. This makes the polynomial have only nonnegative powers of z,

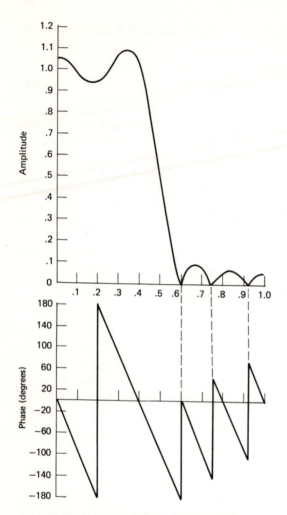

Frequency (in fractions of the Nyquist frequency)

FIG. 3.4.2 The amplitude and phase of the low pass moving average filter with the transfer function of Eq. 3.4.11.

with leading coefficient of unity. The zeros of the transfer function (roots of the equation $H(z) = 0$) have been called $z_1, z_2, \ldots, z_{M-1}$. Since the coefficients $C(I)$ are real, the zeros z_i either occur in complex conjugate pairs or are real themselves (see Exercise 1.2.12).

The factored form of the transfer function provides some insight into the shape of the magnitude and phase characteristics of the filter. Consider first the phase: the phase of a product is the sum of the phases of each factor, by the definition of multiplication. Hence, assuming z is on the unit circle and is equal to $e^{j\omega}$,

$$\text{Arg } H(z) = \text{Arg } C(1) + \text{Arg } z^{-(M-1)} + \text{Arg } (z-z_1) + \ldots + \text{Arg } (z-z_{M-1})$$
$$= \text{Arg } C(1) - (M-1)\omega + \text{Arg } (e^{j\omega}-z_1) + \ldots + \text{Arg } (e^{j\omega}-z_{M-1})$$
$$(3.5.2)$$

The Arg of the coefficient $C(1)$ is either $0°$ or $180°$, depending on whether $C(1)$ is positive or negative. The linear phase term $-(M-1)\omega$ arises from the factor $z^{-(M-1)}$, which represents a delay of $(M-1)$ sample periods. As mentioned above, this factor is brought outside so that the remainder of the transfer function can be written as a polynomial in non-negative powers of z. We are left with the factors

$$(z-z_i) = (e^{j\omega}-z_i), \ i=1,\ldots,M-1 \qquad (3.5.3)$$

This complex number can be interpreted as a vector drawn from the point z_i in the complex plane, which represents the root or zero of the polynomial, to the point $e^{j\omega}$, which lies on the unit circle at an angle determined by the frequency ω. Figure 3.5.1 illustrates this situation.

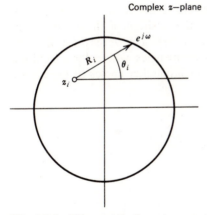

Complex z–plane

FIG. 3.5.1   The contribution of a zero to the transfer function of a moving average digital filter: the factor $(z-z_1)$.

From this diagram we see that the magnitude of $(z-z_i)$, say R_i, is the distance from the zero to the point on the unit circle; and that the angle of $(z-z_i)$ is the angle between the line from z_i to $e^{j\omega}$ and the real axis. Thus, we have the following expression for Arg $H(z)$:

$$\text{Arg } H(z) = \text{Arg } C(1) - (M-1)\omega + \theta_1 + \ldots + \theta_{M-1} \qquad (3.5.4)$$

where the angles θ_i can be read from a graph such as Fig. 3.5.1.

Turning now to the magnitude of $H(z)$, we have

$$
\begin{aligned}
|H(z)| &= |C(1)| \, |z^{-(M-1)}| \, |z-z_1| \ldots |(z-z_{M-1})| \\
&= |C(1)| \, R_1 \ldots R_{M-1}
\end{aligned}
\tag{3.5.5}
$$

Thus, the magnitude of $H(z)$ on the unit circle in the z-plane, which we have seen is the amplitude response, is determined by the distances from any particular point on the circle to all the zeros of the transfer function. If a point on the unit circle corresponding to some frequency F_1 is far from any zeros, the amplitude response at the frequency F_1 will be large. If some other point, corresponding to the frequency F_2, is close to some zeros, the response at F_2 will be small. If in fact a zero lies exactly on the unit circle, the response corresponding to that frequency will be zero.

As an illustration, consider the 10th order digital filter whose transfer function was discussed in the preceding section, and whose amplitude and phase response are plotted in Fig. 3.4.2. Figure 3.5.2 shows the location of the zeros of this transfer function. We note first that there are three zeros precisely on the unit circle in the upper half plane. These correspond to the points in the amplitude plot where the transfer function becomes zero. The three zeros on the unit circle in the lower half plane represent the mirror image of these points for negative ferquencies. Second, we note that there is a close connection between the position of the 10 zeros and the shapes of the magnitude and phase curves. The zeros near the unit circle occur at angles corresponding to frequencies where the magnitude of the transfer function is small. The zeros at smaller angles in the z-plane

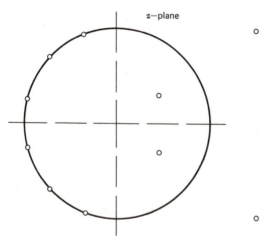

FIG. 3.5.2 Location of the zeros of the filter whose amplitude and phase are shown in Fig. 3.4.2.

occur farther from the unit circle, and their effect is to flatten the magnitude characteristic in the region where it is relatively large. As discussed above, the amplitude and phase curves can be calculated directly from the position of these 10 zeros (plus a knowledge of the multiplicative constant $C(1)$). By experimentation a designer can develop the ability to design approximations to desired frequency characteristics by proper placement of the zeros. There are also more automatic methods of design, which we shall not go into here. The problem of choosing the zeros of a moving average digital filter to achieve a particular prescribed amplitude or phase characteristic has in fact received quite a bit of attention in the literature.

Example Problem

(a) Design a digital filter that has zero output when the input is a phasor of frequency $\pi/4$ radians/sample interval.

(b) Calculate the output of your filter when the following digital filter is applied as an input:

$$X(k) = \begin{cases} 0 & k<0 \\ \sin \dfrac{k\pi}{4} & k\geq 0 \end{cases} \qquad (3.5.6)$$

(c) Explain any discrepancy between the results of parts (a) and (b).

Solution. A phasor at the frequency $\pi/4$ radians/sample interval corresponds to the value $z = e^{j\pi/4}$. Hence, if we put a zero in the filter transfer function at this frequency, the response to a phasor at this frequency will be precisely zero. We should also put in a zero at $z = e^{-j\pi/4}$, so that the zeros occur in complex conjugate pairs, and the coefficients of the filter are real numbers. In accordance with Eq. 3.5.1, then, we can write with $M = 3$:

$$H(z) = C(1) z^{-2} (z - e^{j\pi/4})(z - e^{-j\pi/4}) \qquad (3.5.7)$$

For simplicity, we can take $C(1) = 1$, so that

$$\begin{aligned} H(z) &= z^{-2}(z - e^{j\pi/4})(z - e^{-j\pi/4}) \\ &= z^{-2}(z^2 - 2z \cos \frac{\pi}{4} + 1) \qquad (3.5.8) \\ &= 1 - \sqrt{2}\, z^{-1} + z^{-2} \end{aligned}$$

Thus the defining equation of the filter is

$$Y(k) = X(k) - \sqrt{2}\, X(k-1) + X(k-2) \qquad (3.5.9)$$

This defining equation can be used to calculate the output when the input is the signal $\sin k\pi/4$, and the results are listed below in Eq. 3.5.10:

$$
\begin{array}{llllllll}
k & :0 & 1 & 2 & 3 & 4 & 5 & 6\ldots \\
X(k): & 0 & 1/\sqrt{2} & 1 & 1/\sqrt{2} & 0 & -1/\sqrt{2} & -1\ldots \quad (3.5.10) \\
Y(k): & 0 & 1/\sqrt{2} & 0 & 0 & 0 & 0 & 0\ldots
\end{array}
$$

We see that the first sample value of the output, $Y(1)$, is not zero, although all the other output samples are zero. The reason the output is not zero for every k is that the input is not a true phasor, but is sinusoidal only for nonnegative k. Thus, at the point where the input is "turned on," we detect a nonzero output sample value.

Exercise 3.5.1

What is the effect on a moving average filter of replacing z by z^2 everywhere in its transfer function? More specifically, how are the implementation algorithm, the amplitude response, and the phase response affected?

Exercise 3.5.2

Prove that the magnitude of the transfer function on the unit circle of a moving average filter is not changed (except for a constant factor) if a zero at $z = a$ is replaced by a zero at $z = 1/a$, where a is real. What is the effect of such a change on the phase? Repeat for the case where a pair of complex zeros at $z = a, a^*$ is replaced by the pair $z = 1/a, 1/a^*$.

Exercise 3.5.3 (computer experiment)

Write a computer program to calculate the amplitude and phase of a moving average filter that accepts as input data the location of the zeros in the complex z-plane.

Design a filter that attenuates the amplitude of those phasors whose frequencies are not in the band from $F = .25$ to $.75$ (in fractions of the Nyquist frequency), by experimenting with the location of zeros. Such a filter is called a bandpass filter, since it is designed to pass frequencies only in a specified band. Plot the amplitude and phase of your final filter design.

Exercise 3.5.4

We wish to design a moving average filter H whose transfer function $H(z)$ satisfies the following conditions:

1. H has real coefficients.

2. The value of the transfer function at frequency zero radians/sample interval is $+1$.

3. The value of the transfer function at frequency $\pi/2$ radians/sample interval is $-j$.

Find the transfer function of one such filter. Sketch the amplitude and phase of its frequency response.

Exercise 3.5.5

Show that the amplitude of the transfer function of a moving average filter with real coefficients is an even function of frequency, while the phase is an odd function.

Exercise 3.5.6

Write the transfer functions of the digital filters given in Exercise 3.2.1 in terms of z, and find the locations of their zeros in the complex plane.

Exercise 3.5.7

Find the zeros and sketch the amplitude response of the moving average digital filters with the following transfer functions:

(a) $H(z) = 1 + z^{-1} + z^{-2} + \ldots + z^{-8}$
(b) $H(z) = 1 - z^{-1} + z^{-2} - \ldots + z^{-8}$
(c) $H(z) = 1 + z^{-3} + z^{-6} + z^{-9} + z^{-12}$

Exercise 3.5.8

Find the transfer functions of realizable moving average digital filters with the zero locations shown in the figures below:

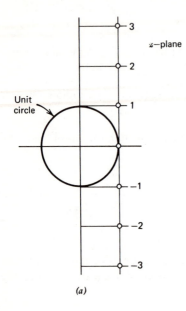

(a)

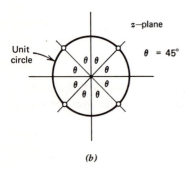

(b)

4.

THE TRANSFORM METHOD

4.1 DEFINITION OF THE z-TRANSFORM

In the preceding chapter we dealt with the situation where a phasor was processed by a moving average digital filter, and we saw that such a filter changed the amplitude and phase angle of the input phasor according to a complex-valued function of the frequency. This approach tells us something about the frequency dependent characteristics of moving average filters, but is quite limited in two important respects: first, it deals only with a phasor input of a single frequency that has been applied for all time. Thus the approach gives information only about the steady-state behavior of the filter when the input is a phasor, but tells us nothing about the response to any other kind of input. Second, the approach is not naturally suited to dealing with filters other than those of the moving average type. We shall want later on to deal with a second type of digital filter, called the recursive filter. For these reasons, we shall develop a more general method for describing the action of a digital filter on a digital signal: the z-transform method.

Why did the phasor approach work the way it did for a moving average filter? Because shifting a phasor by a sample period is equivalent to multiplying it by a complex number; that is,

$$e^{j\omega(k-1)} = e^{-j\omega} e^{j\omega k} \tag{4.1.1}$$

By defining

$$z = e^{j\omega} \tag{4.1.2}$$

we can express the effect of any moving average filter by multiplication by a polynomial in z^{-1}. It is important to realize that this trick works only for phasors and, in fact, these phasors must be defined for all k. Suppose, for example, that a digital signal X is defined to be zero for negative k, and a phasor for nonnegative k, as follows:

$$X(k) = \begin{cases} 0 & k<0 \\ e^{j\omega k} & k\geq 0 \end{cases} \qquad (4.1.3)$$

Now if we shift X to the right, we find that

$$X(k-1) = \begin{cases} 0 & k\leq 0 \\ e^{-j\omega}e^{j\omega k} & k>0 \end{cases} \qquad (4.1.4)$$

From this equation we see that the shifted signal at $k = 0$ is *not* equal to the original signal at $k = 0$ multiplied by $e^{-j\omega}$.

An even more dramatic example is provided by the following signal:

$$X(k) = \begin{cases} 0 & k<0 \\ 1 & k\geq 0 \end{cases} \qquad (4.1.5)$$

which we shall call the *unit step function* (Fig. 4.1.1).

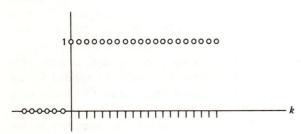

FIG. 4.1.1   Unit step function.

Shifting to the right by one sample period yields the signal

$$X(k-1) = \begin{cases} 0 & k\leq 0 \\ 1 & k>0 \end{cases} \qquad (4.1.6)$$

and this is certainly not equal to $X(k)e^{-j\omega}$; in fact, we cannot even say what ω might be in this case, since X is not a sinusoidal signal.

It would be very convenient if the operation of a shift could be made to correspond to multiplication for, as we have seen, this enables us to represent the action of a moving average filter by multiplication. We shall accomplish this by defining a transformation, called the *z-transformation,* that for every digital signal defines a totally different kind of function, called its *z-transform.* Before we define the *z*-transform, we shall restrict the class of signals under consideration to those that are zero for negative k. Such signals are called *one-sided* signals. This restriction could be removed, at the expense of more complicated mathematics, but we shall not find that necessary here. Thus we shall deal with signals that are "turned on" at $k = 0$, which is in contrast with the phasor approach, where all signals were applied for all k. We can now define the z-transform:

Definition

Let X (k) be a digital signal that is zero for $k < 0$. Its z-transform, denoted by X^* (z), is defined to be the function of z:

$$X^*(z) = X(0) + X(1)z^{-1} + X(2)z^{-2} + \ldots$$
$$= \sum_{k=0}^{\infty} X(k)z^{-k} \tag{4.1.7}$$

A number of comments about this definition are now in order.

The first question is: what is z? We shall see in the next section that because of the shift-multiplication property that we are aiming for, z will have the same interpretation as it did in the phasor approach. That is, when z is on the unit circle in the complex z-plane, its angle is interpreted as a frequency variable. But the question is not how we interpret z, but rather how we define z. The answer is: z is an independent complex variable. It has much the same status as k, the sample number. A signal is defined to be a function of k, while its z-transform is defined to be a function of z. Thus, z is the *domain* of the z-transform of a signal. We speak, in fact, of the z-transform as a transformation from the *time domain* to the *frequency domain*. We shall use the following symbolism for the z-transform:

$$F(k) \xrightarrow{Z} F^*(z) \tag{4.1.8}$$

The symbol over the arrow indicates the name of the transformation.

We can note at this point that a digital filter can be thought of as a transformation in the same way as the z-transform. For example, if X is the input to a filter H, and Y is the output, we write

$$X(k) \xrightarrow{H} Y(k) \tag{4.1.9}$$

Thus, a filter is a transformation that converts one function of k to another function of k, while the z-transform converts one function of k to another function of z.

The next question that might logically arise about the z-transform concerns the convergence of the defining power series (4.1.7). In general, $X(k)$ will be nonzero for an infinite number of values of k, so that this series will, in fact, be infinite. We then must think about the convergence of this series for various values of z. The series is a power series in *negative* powers of z, and for most digital signals it will converge if the modulus of z is sufficiently *large*. We can use the following mathematical fact about series: if a power series in negative powers of z converges for some value of z, say z_1, then it will also converge for all values of z that have a modulus larger than $|z_1|$. Intuitively, this can be seen by examining the modulus of the kth term, which is

$$\left|\frac{X(k)}{z^k}\right| = \frac{|X(k)|}{|z|^k} \qquad (4.1.10)$$

If we increase the modulus of z, we decrease the modulus of every term and, hence, will tend to help the convergence of the series. Thus, if we establish one value z_1 for which the series converges, we then know that the series converges for all values of z outside the circle with center at $z = 0$ and radius extending to the point z_1 (Fig. 4.1.2.).

Usually, we shall be interested only in the functional form of the z-transform, and the question of convergence will not have a direct effect on what we are doing. Convergence questions do become important in more advanced work, however, especially where signals that are not one-sided are considered.

In the next section we consider some important properties of z-transforms.

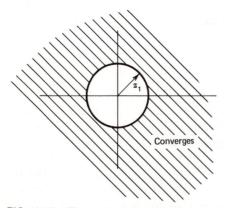

FIG. 4.1.2   The region of convergence of a z-transform that is guaranteed when we know that the z-transform converges for the value $z = z_1$.

Exercise 4.1.1

Consider a moving average filter whose input is the sum of two phasors of different frequency. Is the output a complex number times the input?

Exercise 4.1.2

What digital signal $X(k)$ has the z-transform $X^*(z) = 1$?

4.2 SOME PROPERTIES OF z-TRANSFORMS

We can now describe precisely the property that we wanted to carry over from the phasor approach:

Property 1

Let $X(k)$ be a digital signal that is zero for $k < 0$. Then if

$$X(k) \xrightarrow{\quad Z \quad} X^*(z) \tag{4.2.1}$$

then

$$X(k-1) \xrightarrow{\quad Z \quad} z^{-1}X^*(z) \tag{4.2.2}$$

Proof

From the definition of the z-transform

$$X^*(z) = \sum_{k=0}^{\infty} X(k) z^{-k} \tag{4.2.3}$$

Multiplying both sides by z^{-1}, we obtain

$$z^{-1}X^*(z) = \sum_{k=0}^{\infty} X(k) z^{-(k+1)} \tag{4.2.4}$$

Now replace the index of summation on the right by

$$k = n - 1 \qquad (4.2.5)$$

yielding

$$z^{-1} X^* (z) = \sum_{n=1}^{\infty} X(n-1) z^{-n} \qquad (4.2.6)$$

The index n starts from $n = 1$ since $n = k + 1$, and k started from 0. We may nevertheless start the summation from $n = 0$, since we have demanded that $X(-1) = 0$. Hence

$$z^{-1} X^* (z) = \sum_{n=0}^{\infty} X(n-1) z^{-n} \qquad (4.2.7)$$

which is just another way of writing

$$X(n-1) \xrightarrow{\quad Z \quad} z^{-1} X^* (z)$$

This is the same as (4.2.2), since n on the left-hand side is an independent variable representing the sample number, and can be replaced by k.

This property is the fundamental reason why z-transforms are useful. It allows us to represent moving average filters and, as we shall see, other linear time-invariant filters, by multiplication, in much the same way as in the phasor approach.

The next property is again crucial for our purposes. It expresses the fact that the z-transform of a sum is the sum of z-transforms, and allows us to treat terms in a filter equation one at a time.

Property 2 (linearity of the z-transform)

Let $X(k)$ and $Y(k)$ be two digital signals that are zero for $k < 0$. Then if

$$X(k) \xrightarrow{\quad Z \quad} X^* (z) \qquad (4.2.8)$$

$$Y(k) \xrightarrow{\quad Z \quad} Y^* (k) \qquad (4.2.9)$$

and if a and b are any two constants, then

$$aX(k) + bY(k) \xrightarrow{\quad Z \quad} aX^*(z) + bY^*(k) \tag{4.2.10}$$

Proof

From the definition of the z-transform

$$\sum_{k=0}^{\infty} [aX(k) + bY(k)] z^{-k} = a \sum_{k=0}^{\infty} X(k) z^{-k} + b \sum_{k=0}^{\infty} Y(k) z^{-k} = aX^*(z) + bY^*(z) \tag{4.2.11}$$

which is equivalent to (4.2.10).

The last two properties we shall derive in this section will be of use to us in deriving a useful collection of z-transforms. The first of the two concerns the z-transform of the signal obtained by multiplying a given signal by the factor a^k at each sample instant k.

Property 3

Let $X(k)$ be a digital signal that is zero for $k < 0$, and that has the z-transform $X^*(z)$. Then if a is any constant

$$a^k X(k) \xrightarrow{\quad Z \quad} X^*(a^{-1}z) \tag{4.2.12}$$

Proof

Again, from the definition of the z-transform

$$\sum_{k=0}^{\infty} a^k X(k) z^{-k} = \sum_{k=0}^{\infty} X(k) (a^{-1}z)^{-k} = X^*(a^{-1}z) \tag{4.2.13}$$

The last property concerns the z-transform of the signal $kX(k)$.

Property 4

Let $X(k)$ be a digital signal that is zero for negative k, and with z-transform $X^*(z)$. Then

$$kX(k) \xrightarrow{Z} -z\frac{d}{dz}[X^*(z)] \qquad (4.2.14)$$

Proof

We calculate the derivative of $X^*(z)$ as follows:

$$\frac{d}{dz}[X^*(z)] = \frac{d}{dz}\sum_{k=0}^{\infty} X(k)z^{-k} = \sum_{k=1}^{\infty}(-k)X(k)z^{-k-1} \qquad (4.2.15)$$

The last summation starts at $k = 1$, since the derivative of the $k = 0$ term is the derivative of a constant, which is zero. We can justify differentiating the infinite series term by term in this case within its region of convergence because of the mathematical properties of power series. Now the factor k in the last summation allows us to start the summation at $k = 0$ after all, since the $k = 0$ term simply adds zero to the summation. Hence

$$\frac{d}{dz}[X^*(z)] = -z^{-1}\sum_{k=0}^{\infty} kX(k)z^{-k} \qquad (4.2.16)$$

bringing the constant $(-z^{-1})$ outside the summation. Finally, multiplying by $(-z)$ we obtain

$$(-z)\frac{d}{dz}[X^*(z)] = \sum_{k=0}^{\infty} kX(k)z^{-k} \qquad (4.2.17)$$

which is the same as (4.2.14).

Exercise 4.2.1

Find the z-transform of the digital signal

$$X(k) = \begin{cases} 1/k! & k \geq 0 \\ 0 & k < 0 \end{cases} \qquad (4.2.18)$$

Exercise 4.2.2

Derive a formula similar to (4.2.14) by integrating the equation defining the z-transform term by term.

4.3 MOVING AVERAGE FILTERS REVISITED

We can now use Properties 1 and 2 to show that the transfer function derived in the phasor approach is meaningful in the z-transform approach as well and, in fact, has a more general interpretation. Consider as we did in Chapter 3 the digital filter H that has input signal X and output signal Y:

$$X(k) \xrightarrow{\quad H \quad} Y(k) \tag{4.3.1}$$

and is a moving average filter of order $M-1$, with coefficients $C(i)$, so that the defining equation is

$$Y(k) = C(1)X(k) + C(2)X(k-1) + \ldots + C(M)X(k-M+1) \tag{4.3.2}$$

By Property 2, the linearity of the z-transform, the z-transform of this sum is equal to the sum of the z-transforms of each term. By Property 1, the z-transform of each term can be written as a multiple of the z-transform of $X(k)$ as follows:

$$C(1)X(k) \xrightarrow{\quad Z \quad} C(1)X^*(z)$$

$$C(2)X(k-1) \xrightarrow{\quad Z \quad} C(2)z^{-1}X^*(z)$$

$$\vdots$$

$$C(M)X(k-M+1) \xrightarrow{\quad Z \quad} C(M)z^{-(M-1)}X^*(z) \tag{4.3.3}$$

Hence,

$$Y(k) \xrightarrow{\quad Z \quad}$$
$$Y^*(z) = [C(1) + C(2)z^{-1} + \ldots + C(M)z^{-(M-1)}]X^*(z) \tag{4.3.4}$$

We see then that the transfer function

$$H(z) = C(1) + C(2)z^{-1} + \ldots + C(M)z^{-(M-1)} \tag{4.3.5}$$

is equal to the ratio of the z-transform of the output to the z-transform of the input; that is,

$$H(z) = \frac{Y^*(z)}{X^*(z)} \tag{4.3.6}$$

We sometimes express this relationship by the diagram in Fig. 4.3.1.

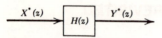

$X^*(z) \longrightarrow$ $\boxed{H(z)}$ $\longrightarrow Y^*(z)$

FIG. 4.3.1 Representation of the multiplicative effect of the transfer function on the z-transform of the input signal.

Now it is important to realize the more general nature of this relationship (4.3.6) when compared with the similar result in the phasor case. It is still true enough that when X is a phasor, then Y is also, and the *ratio of the phasors* Y/X is given by $H(z)$ when z is equal to $e^{j\omega}$ (i.e., when z is a complex variable on the unit circle). But (4.3.6) says something quite different and more generally applicable: if $X(k)$ is *any one-sided* digital signal, then the z-transform of the output signal can be obtained simply by multiplying the z-transform of the intput signal by the transfer function $H(z)$. This result allows us to deal with the situation where the input to a moving average filter is an arbitrary one-sided digital signal.

The phasor approach provides us with useful information, nonetheless. It tells us that we can interpret the transfer function $H(z)$ as the frequency response at the frequency ω when it is evaluated at the value of z on the unit circle $z = e^{j\omega}$.

Let us now consider a simple example of the preceding result. Let the input signal $X(k)$ be defined as follows:

$$
\begin{aligned}
X(k) &= 0 \qquad k<0 \\
X(0) &= 1 \\
X(1) &= 1 \\
X(k) &= 0 \qquad k>1
\end{aligned}
\tag{4.3.7}
$$

The z-transform of X is, from the definition of the z-transform,

$$X^*(z) = 1 + z^{-1} \tag{4.3.8}$$

The input signal X can be represented as in Fig. 4.3.2.

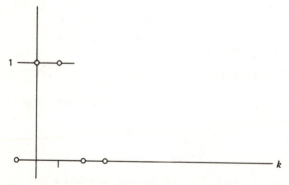

FIG. 4.3.2 The digital signal of Eq. 4.3.7.

Now suppose that X is applied to the moving average filter defined by

$$Y(k) = X(k) + X(k-1) \qquad (4.3.9)$$

which is the simple filter used as an example in Chapter 3. The transfer function of H is

$$H(z) = 1 + z^{-1} \qquad (4.3.10)$$

Hence

$$\begin{aligned}
Y^*(z) &= H(z)X^*(z) \\
&= (1 + z^{-1})(1 + z^{-1}) \\
&= 1 + 2z^{-1} + z^{-2}
\end{aligned} \qquad (4.3.11)$$

This z-transform of Y is in the form of a power series in decreasing powers of z, and so we may interpret the coefficients as values of the signal $Y(k)$. Hence

$$\begin{aligned}
Y(k) &= 0 \qquad k < 0 \\
Y(0) &= 1 \\
Y(1) &= 2 \\
Y(2) &= 1 \\
Y(k) &= 0 \qquad k > 2
\end{aligned} \qquad (4.3.12)$$

Thus, we have determined the sample values of the output by finding its z-transform, and then by determining what signal Y has this function as its z-transform. We may interpret this last step as the process of taking the *inverse* z-transform. The result (4.3.12) may be checked from the defining equation of the filter (4.3.9) as follows:

$$\begin{aligned}
Y(-1) &= X(-1) + X(-2) = 0 \\
Y(0) &= X(0) + X(-1) = 1 \\
Y(1) &= X(1) + X(0) = 2 \\
Y(2) &= X(2) + X(1) = 1 \\
&\ldots \text{and so on}
\end{aligned} \qquad (4.3.13)$$

The whole situation can be depicted by the diagram in Fig. 4.3.3, which illustrates the fact that the filtering operation is represented in the z-transform domain by multiplication by the appropriate transfer function. We can interpret the steps in the example above as starting from $X(k)$ in the diagram, moving to the right, down, and to the left; instead of just moving down.

We shall next calculate the z-transforms of some commonly encountered and useful signals. Then we shall give more attention to the problem of taking the inverse z-transform.

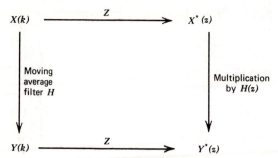

FIG. 4.3.3   Diagram illustrating that filtering in the time domain corresponds to multiplication in the z-transform domain.

Example Problem

Prove that the inverse z-transform is unique; that is, if two signals have the same z-transform, they are identical.

Solution. Let $X_1(k)$ and $X_2(k)$ be two digital signals with the same z-transform. That is,

$$\sum_{k=0}^{\infty} X_1(k) z^{-k} = \sum_{k=0}^{\infty} X_2(k) z^{-k}$$

Subtracting the right-hand side from both sides, this becomes

$$\sum_{k=0}^{\infty} X_1(k) z^{-k} - \sum_{k=0}^{\infty} X_2(k) z^{-k} = 0$$

Subtracting the series term by term, as we may do for power series within the region of convergence, we get

$$\sum_{k=0}^{\infty} [X_1(k) - X_2(k)] z^{-k} = 0$$

For the sum of a power series to be identically zero, it must be that each term is zero (this is a property of power series). Hence

$$X_1(k) = X_2(k) \qquad \text{for every } k$$

which is what we wished to prove.

This property of z-transforms enables us to move freely from the time-domain to the frequency domain, and back, as in Fig. 4.3.3, with assurance that every step yields a unique result.

Exercise 4.3.1

Consider the input signal defined below:

$$X(k) = \begin{cases} 1 & 0 \leq k \leq 10 \\ 0 & \text{otherwise} \end{cases}$$

Calculate the output signal $Y(k)$ for all k in two ways; first, using the filtering equation and, second, using the z-transform method, for the following moving average filters:

(a) $H(z) = 1 - z^{-1}$

(b) $H(z) = 1 + z^{-1}$

(c) $H(z) = (1 - z^{-1})^2$

Exercise 4.3.2

Repeat the exercise above for the input signal

$$X(k) = \begin{cases} 0 & k < 0 \\ 1 & k \geq 0 \end{cases}$$

Exercise 4.3.3

The logarithm can be thought of as a transformation of real numbers that converts multiplication to addition, just as the z-transform converts filtering to multiplication. Draw a diagram similar to that in Fig. 4.3.3 with the left-right transformation being the logarithm instead of the z-transform.

4.4 A COLLECTION OF z-TRANSFORMS

The simplest kind of z-transform arises when the signal under consideration is zero at all but a finite number of sample numbers. In this case the definition of the z-transform leads simply to a finite polynomial in z^{-1}, with the coefficient of each term z^{-k} being simply the value of the signal at the kth sample number. As an example, take the signal

$$X(k) = \begin{cases} 0 & k \leq 0 \\ .5 & k = 1 \\ 1 & k = 2 \\ .5 & k = 3 \\ 0 & k > 3 \end{cases} \qquad (4.4.1)$$

By definition, the z-transform of X is

$$X^*(z) = .5z^{-1} + z^{-2} + .5z^{-3} \qquad (4.4.2)$$

The inverse z-transform problem is trivial here as well, since the sample values are obtained from the z-transform simply by reading the coefficients of the powers of z.

The more interesting case occurs when the signal is nonzero for an infinite number of values of k; in other words, the signal goes on forever. Of course, such signals exist only in our imagination, since all physical processes of use to us in processing information terminate in a finite amount of time. However, when the signal goes on for thousands, or even millions of sample values, the concept of a signal of infinite extent is very useful. We shall deal here with a particularly important class of such signals: those that are exponential or sinusoidal in nature. The z-transforms of all these signals will be derived from one simple starting point using the properties derived in Section 4.2. We shall assume throughout that all signals are one-sided.

The simple starting point is provided by the unit step function $X(k) = 1$ discussed before, and plotted in Fig. 4.1.1. The z-transform of this signal is by definition

$$X^*(z) = 1 + z^{-1} + z^{-2} + z^{-3} + \ldots \qquad (4.4.3)$$

This infinite series is a geometric one, and we know from algebra that it is equal to (see Chapter 1)

$$X^*(z) = \frac{1}{1 - z^{-1}} \qquad (4.4.4)$$

provided that the ratio of sequential terms is less in magnitude than 1. We write the condition for convergence as

$$|z^{-1}| < 1 \qquad (4.4.5)$$

Since

$$|z^{-1}| = \frac{1}{|z|} \qquad (4.4.6)$$

This is equivalent to

$$|z| > 1 \qquad (4.4.7)$$

Now for the first time we encounter a z-transform that is *not* a finite polynomial in z^{-1}. Let us examine it more closely. If we multiply numerator and denominator by z, (4.4.4) becomes

$$X^*(z) = \frac{z}{z-1} \qquad (4.4.8)$$

A new phenomenon occurs here: when $z = 1$ this z-transform becomes infinite. (In the finite polynomial case, the only value of z for which a z-transform could become infinite was $z = 0$, and this occurred because we defined the z-transform as a series in z^{-1}.) The point at which the z-transform becomes infinite is called a *pole* and, as we shall see, the poles of a z-transform tell us a great deal about the nature of the signal. Figure 4.4.1 shows the location of the poles and zeros of this transform.

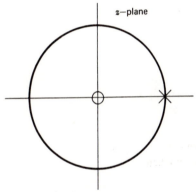

FIG. 4.4.1   Pole and zero of the z-transform of the unit step function, Eq. 4.4.4.

The fact that the pole occurs on the unit circle has a special significance in light of our interpretation of the unit circle as representing the frequency line. Specifically, it means that the signal has an infinite frequency component at zero frequency. This is consistent with our discussion of phasors, where we say that a phasor of zero frequency is a constant, since the step function is constant for an infinite number of sample numbers.

We can find the magnitude of the z-transform when z is on the unit circle, and this quantity is called the *frequency content* of the signal. Let us perform this calculation for the unit step function. Letting $z = e^{j\omega}$, the z-transform becomes

$$X^*(z) = \frac{e^{j\omega}}{1 - e^{j\omega}}$$ (4.4.9)

Taking the magnitude, we obtain

$$|X^*(z)| = \frac{|e^{j\omega}|}{|1 - e^{j\omega}|}$$

$$= \frac{1}{|e^{-j\omega/2} - e^{j\omega/2}|}$$ (4.4.10)

$$= \frac{1}{2|\sin \omega/2|}$$

A sketch of this as a function of the frequency ω is shown in Fig. 4.4.2.

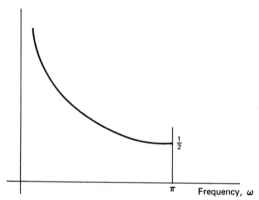

FIG. 4.4.2   Frequency content of the unit step function.

The following question arises: why is there nonzero frequency content at any frequency besides 0? The answer is that the signal is not a constant for all k, but is turned on at $k = 0$. This abrupt jump causes all the other frequencies to be present, but the only frequency with an infinite component is the frequency zero. The abrupt starting point at $k = 0$ will cause all the signals under consideration to have frequency components spread throughout the range from zero to the Nyquist frequency.

As we have seen, when a signal is filtered by a moving average filter, its z-transform is multiplied by the transfer function of the filter. Since the magnitude of a product is the product of the magnitudes, this implies that the frequency content of the output signal can be obtained by multiplying the frequency content of the input by the magnitude of the transfer function on the unit circle. This frequency domain interpretation of filtering is our primary reason for studying z-transforms. It allows us to think of linear time-invariant filtering in terms of frequency content.

Now if we let c be any real number, and apply Property 3, we obtain the following transform

$$c^k \quad k \geq 0 \xrightarrow{\quad Z \quad} \frac{1}{1 - cz^{-1}} \qquad (4.4.11)$$

When c is between 0 and 1 this signal has a decreasing exponential shape (see Fig. 4.4.3).

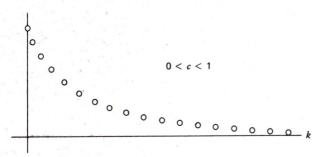

FIG. 4.4.3   The digital signal c^k, $0 < c < 1$.

When the magnitude of a signal decreases to zero as k becomes infinite, such as occurs here, we say the signal is *damped*. When c is between -1 and 0, the signal is damped but alternates in sign, as in Fig. 4.4.4.

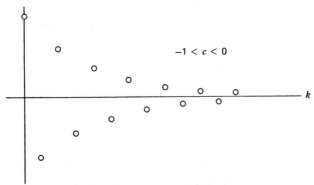

FIG. 4.4.4   The digital signal c^k, $-1 < c < 0$.

When c is greater than 1, the signal grows without bound in one direction; and when c is less than -1 the signal grows without bound but alternates in sign (see Figs. 4.4.5 and 4.4.6).

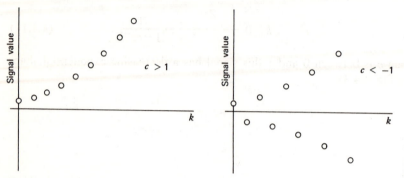

FIG. 4.4.5   The digital signal c^k, $c>1$.   FIG. 4.4.6   The digital signal c^k, $c<-1$.

The pole-zero plot of our exponential signal is shown in Fig. 4.4.7.

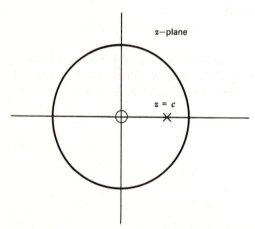

FIG. 4.4.7   Pole and zero of the z-transform of the exponential digital signal c^k, Eq. 4.4.11.

The pole occurs at the point $z = c$, and determines the behavior of the function as k grows large. A pole *outside* the unit circle is associated with *unbounded* growth; while a pole inside the unit circle is associated with damped behavior. A pole directly on the unit circle, as in the case of a unit step, is associated with borderline behavior in this respect; either taking on a constant value as k grows large ($c = 1$), or remaining constant in magnitude but alternating in sign ($c = -1$).

The significance of the pole location in determining the behavior of a signal as k becomes large is quite general: any pole outside the unit circle will imply that the signal becomes unbounded as k gets large. We call such a signal *unstable*, and we shall usually restrict ourselves to *stable* signals. Unstable signals will tend to produce overflow in a computer. (Unstable analog signals will cause saturation and other nonlinear effects in continuous-time equipment and are usually undesirable for these analogous reasons.)

We shall now calculate the z-transforms of sinusoidal signals by considering the case where c is complex. If we let

$$c = ae^{jb} \qquad (4.4.12)$$

we have the z-transform (4.4.11) already calculated:

$$a^k e^{jkb} \quad k \geq 0 \xrightarrow{\quad Z \quad} \frac{1}{1 - ae^{jb}z^{-1}} \qquad (4.4.13)$$

If we take the real and imaginary parts of this z-transform, we obtain two z-transforms as follows:

$$a^k \cos kb \quad k \geq 0 \xrightarrow{\quad Z \quad} \text{Real}\left(\frac{1}{1 - ae^{jb}z^{-1}}\right) = \frac{1 - a \cos b \, z^{-1}}{1 - 2a \cos b \, z^{-1} + a^2 z^{-2}}$$
$$(4.4.14)$$

$$a^k \sin kb \quad k \geq 0 \xrightarrow{\quad Z \quad} \text{Imag}\left(\frac{1}{1 - ae^{jb}z^{-1}}\right) = \frac{a \sin b \, z^{-1}}{1 - 2a \cos b \, z^{-1} + a^2 z^{-2}}$$
$$(4.4.15)$$

The poles of these transforms occur at the points $z = ae^{\pm jb}$; that is, at points at angles $\pm b$ at a distance a from the origin. The poles and zeros of the signal $a^k \cos kb$ are shown in Fig. 4.4.8.

Notice that when a pole is near the unit circle, the frequency content of the signal tends to be large at frequencies on the unit circle near the pole. This is opposite to the effect of a zero being near the unit circle, which we observed in studying the transfer functions of moving average filters.

Applying Property 4 to the z-transform of the unit step yields the following z-transform:

$$X(k) = k \quad k \geq 0 \xrightarrow{\quad Z \quad} -z \frac{d}{dz}\left(\frac{1}{1 - z^{-1}}\right) = \frac{z^{-1}}{(1 - z^{-1})^2}$$
$$(4.4.16)$$

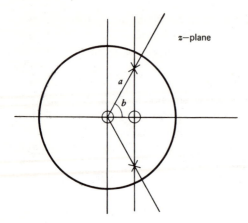

FIG. 4.4.8   Poles and zeros of the sinusoidal digital signal $a^k \cos kb$, Eq. 4.4.14.

This is the transform of a linearly increasing signal called a *ramp* (Fig. 4.4.9).

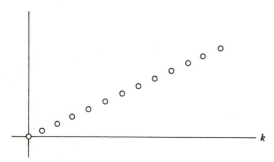

FIG. 4.4.9   The digital ramp signal, $X(k) = k$.

This transform has *two* poles exactly on the unit circle at the same point: it is even more on the borderline of stability in that respect than the step signal. It becomes unbounded, but its rate of growth is slower than exponential. By repeated application of Property 4 we could derive the z-transforms of the signals

$$X(k) = k^n \qquad k \geq 0 \qquad\qquad (4.4.17)$$

where n is any positive integer.

We have now derived all the z-transforms we shall need in usual applications. Figure 4.4.10 provides a table of the results. The transforms all have the property that they are ratios of polynomials in z. These functions are called *rational* functions and are really the only kind of z-transforms we need consider.

$F(k)$	$F^*(z)$
1	$\dfrac{1}{1-z^{-1}}$
c^k	$\dfrac{1}{1-cz^{-1}}$
k	$\dfrac{z^{-1}}{(1-z^{-1})^2}$
$a^k \cos kb$	$\dfrac{1-a\cos bz^{-1}}{1-2a\cos bz^{-1}+a^2z^{-2}}$
$a^k \sin kb$	$\dfrac{a\sin bz^{-1}}{1-2a\cos bz^{-1}+a^2z^{-2}}$

FIG. 4.4.10   Table of commonly used z-transforms.

Exercise 4.4.1

Find the z-transforms and plot the pole-zero patterns of the following signals:

(a) $1 + k$
(b) $c^k + c^{-k}$ c real
(c) $(-1)^k (2)^{-k}$
(d) kc^k c real
(e) $ka^k \cos kb$
(f) k^2
(g) $(-1)^k$

Exercise 4.4.2

Check the calculations of (4.4.14) and (4.4.15).

Exercise 4.4.3

Plot the frequency content of

(a) c^k

(b) k

Exercise 4.4.4

Find an algebraic expression for the frequency content of $a^k \cos kb$.

Exercise 4.4.5

Find the region of convergence for the z-transforms given in Fig. 4.4.10.

4.5 INVERSE z-TRANSFORM, METHOD 1: LONG DIVISION

We turn now to the problem of calculating the inverse z-transform. That is, we assume we are given a ratio of polynomials in z^{-1}, and we wish to find the signal that has this function of z as its z-transform. Of course, the simplest method would be to find the rational z-transform in the table of Fig. 4.4.10. (Extensive tables appear in the references at the end of Chapter 6.) We shall often deal with rational functions that are not in this table, however, since the transforms we shall encounter will have numerators and denominators of degree in z^{-1} higher than 2. If the inverse z-transform we are looking for represents the output of a digital filter, we can of course use the computer to implement the digital filter, supplying the appropriate input signal. The output of the filter will then be the desired inverse z-transform. The method described in this section has the advantage of being quick and easy to use with pencil and paper. It has the additional advantage of requiring very little storage if programmed for the computer. However, it gives only numerical results and does not provide insight into the composition of the output signal. (In the next chapter we shall describe an inverse z-transform method that gives the output signal as a sum of the simple exponential and sinusoidal signals discussed above.)

The method is simply the process of long division of two polynomials. Take, for an example, the rational transform

$$Y^*(z) = \frac{.5z^{-1}}{1 + .25z^{-2}} \qquad (4.5.1)$$

Dividing out we obtain

$$
\begin{array}{r}
.5z^{-1} - .125z^{-3} + .03125z^{-5} - .0078125z^{-7} \ldots \\
1 + .25z^{-2} \overline{)\ .5z^{-1}} \\
\underline{.5z^{-1} + .125z^{-3}} \\
-.125z^{-3} \\
\underline{-.125z^{-3} - .03125z^{-5}} \\
+.03125z^{-5} \\
\underline{.03125z^{-5} + .0078125z^{-7}} \\
-.0078125z^{-7} \\
\ldots
\end{array}
\qquad (4.5.2)
$$

Thus the sample values of $Y(k)$ are

$$
\begin{aligned}
Y(0) &= 0.0 \\
Y(1) &= 0.5 \\
Y(2) &= 0.0 \\
Y(3) &= -0.125 \\
Y(4) &= 0.0 \\
Y(5) &= 0.03125 \\
Y(6) &= 0.0 \\
Y(7) &= -.0078125 \ldots
\end{aligned}
\qquad (4.5.3)
$$

Inspection of the table in Fig. 4.4.10 shows that this transform corresponds to the signal

$$Y(k) = a^k \sin kb \qquad (4.5.4)$$

with $a = .5$ and $b = \pi/2$. The values obtained for $Y(k)$ can be checked against $(.5)^k \sin k\pi/2$.

Exercise 4.5.1

Calculate the first 5 nonzero sample values of the signals with the following z-transforms, using long division:

(a) $1/(1-z^{-1})^2$ (c) $.3z^{-1}/(1-.8z^{-1}+.25z^{-2})$
(b) $1/(1-z^{-1})^3$ (d) $(1+.5z^{-1})/(1-3.5z^{-1}+1.5z^{-2})$
                              (e) $1/(1-z^{-1}+.5z^{-2})$

Exercise 4.5.2 (computer experiment)

Write a FORTRAN program that calculates the inverse z-transform of the rational function

$$\frac{C(1) + C(2)z^{-1} + C(3)z^{-2} + \ldots + C(M)z^{-(M-1)}}{1 + D(1)z^{-1} + D(2)z^{-2} + \ldots + D(L)z^{-L}} \qquad (4.5.5)$$

The following steps should be performed:

(1) Read in M, the number of numerator coefficients; L, the number of denominator coefficients; and NPTS, the number of output points desired. We assume that M and L are each at most 10.

(2) Read the coefficients C and D into arrays dimensioned 10.

(3) Define a vector ROW of dimension 11 that is to represent the result under the line in each step of the long division. Initialize it to the numerator coefficients. Be sure to set the remainder of ROW to zero.

(4) Write out the first inverse z-transform sample value, which is simply $C(1) = ROW(1)$.

(5) Calculate the next set of values in ROW using the last sample value, the last values in ROW, and the denominator coefficients. If Y is the last sample value, then the new values in ROW are calculated by

$$ROW(I) = ROW(I+1) - Y * D(I), \quad I = 1, \ldots, 10 \qquad (4.5.6)$$

(6) Write out the new sample value, which is simply $ROW(1)$, and go back to 5, until the required number of sample values have been written out. For debugging purposes, it may be desirable to print out the entire ROW vector at each stage.

Notice that the total storage required for the arrays is just 31, assuming maximum values of M and L of 10. We need not save the values of the inverse transform.

Put in the following feature: test the value of the inverse z-transform at each stage, and if its magnitude exceeds some very large number, or is not zero and below some very small number, terminate the calculation. This will prevent overflow if the signal is unstable, or underflow.

Test the program with various rational functions, including those in Exercise 4.5.1. Case (d) is unstable, so be careful of overflow.

Exercise 4.5.3

The length of the row vector as calculated in (4.5.6) need not necessarily be 10. What is the smallest length it can be, in terms of M and L?

Exercise 4.5.4

What happens if the denominator of a rational z-transform is a factor of the numerator? That is, if the denominator divides the numerator evenly? Test the program written in Exercise 4.5.2 with such an example.

Exercise 4.5.5 (computer experiment)

If we take the inverse z-transform of

$$\frac{\sin b\, z^{-1}}{1 - 2 \cos b\, z^{-1} + z^{-2}} \tag{4.5.7}$$

we should come back to the digital signal $\sin kb$. If we use long division, round-off errors will accumulate until the results are unusable. Experiment with this using the computer. Can you think of a remedy for the build-up of round-off error?

Exercise 4.5.6

If we take the z-transform of a complex-valued digital signal, we will in general obtain a z-transform with complex coefficients. Will long division work in this case?

5.
RECURSIVE
DIGITAL FILTERS

5.1 A SIMPLE RECURSIVE FILTER

One of the reasons for studying the z-transform is to analyze filters other than those of the moving average type. To see how such filters come about, consider the computer implementation of a simple first order moving average filter:

$$\begin{array}{l} \text{DO 10} \quad \text{I}=2,\text{N} \\ \text{10 Y(I)} = \text{X(I)} + .5 * \text{X(I}-1) \end{array} \qquad (5.1.1)$$

In computing the Ith sample value of the output Y, we make use of the Ith sample value of the input X, and the preceding value of $X(I-1)$. Since the preceding value of the output Y has already been calculated, it might occur to us to make use of it instead, as follows:

$$\begin{array}{l} \text{Y(1)} = 0. \\ \text{DO 10} \quad \text{I}=2,\text{N} \\ \text{10 Y(I)} = \text{X(I)} + .5 * \text{Y(I}-1) \end{array} \qquad (5.1.2)$$

We must in this case define the value $Y(1)$ in order to start the filtering process. This value is an example of an *initial condition* of a filter, and is usually taken to be zero unless there is some reason to do otherwise. The initial conditions of a digital filter of this kind are analogous to stored energy at $t = 0$ in a continuous-time filter, such as initial charge on a capacitor or initial current in an inductor. A filter that uses previous values of the *output,* as well as present and previous values of the input, is called *recursive.*

Let us pursue this example of a recursive filter, calling it H. Write the defining equation as

$$Y(k) = X(k) + .5Y(k-1) \qquad (5.1.3)$$

Now this can be thought of as equality between two signals, and if two signals are equal, their z-transforms must be equal. Taking the z-transform of both sides of the equation yields

$$Y^*(z) = X^*(z) + .5z^{-1}Y^*(z) \qquad (5.1.4)$$

where we have used Properties 1 and 2 of the z-transform. Notice that the z-transform has converted the filter equation (5.1.3), which is an example of a *difference* equation, into an *algebraic* equation that can be solved for $Y^*(z)$, the z-transform of the output. Thus, solving for $Y^*(z)$, we obtain

$$Y^*(z) = \frac{1}{1 - .5z^{-1}} X^*(z) \qquad (5.1.5)$$

The function

$$H(z) = \frac{1}{1 - .5z^{-1}} \qquad (5.1.6)$$

plays the same role as the polynomial in z^{-1} in the case of the moving average filters: it is the ratio of the output z-transform to the input z-transform. Hence, the frequency content of the output is obtained by multiplying the frequency content of the input by the values of $H(z)$ on the unit circle in the z-plane. For a recursive filter, we see that the transfer function is not a polynomial in z^{-1}, but rather a *rational function* of z^{-1}. This is the key to the usefulness of recursive filters; we can obtain frequency response shapes with a rational function of z that are not obtainable with a simple polynomial.

Now let us work through an example to show how the output of this digital filter can be computed when the z-transform of the input is known. Suppose that the input X is a unit step function, so that

$$X^*(z) = \frac{1}{1 - z^{-1}} \qquad (5.1.7)$$

Then the output has the z-transform

$$Y^*(z) = \frac{1}{(1 - z^{-1})(1 - .5z^{-1})} = \frac{1}{1 - 1.5z^{-1} + .5z^{-2}} \qquad (5.1.8)$$

Long division yields the following values for the output samples:

$$Y(0) = 1$$
$$Y(1) = 1.5$$
$$Y(2) = 1.75 \qquad (5.1.9)$$
$$Y(3) = 1.875$$
$$Y(4) = 1.9375 \ldots$$

These values can be checked by using the defining filter equation (5.1.3), assuming the initial value $Y(-1) = 0$. Figure 5.1.1 suggests that the output signal approaches the value 2 with an exponential shape. We shall see later that this is true, and that this behavior is determined by the pole locations of the output signal, which are, after all, the combined pole locations of the input signal and the filter transfer function.

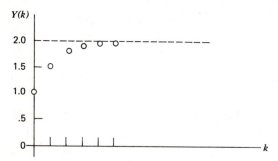

FIG. 5.1.1   Output signal $Y(k)$ in the example.

Example Problem

Show that the recursive digital filter discussed in Section 5.1 is linear and time-invariant.

Solution. The equation that defines the digital filter is

$$Y(k) = X(k) + .5Y(k-1)$$

Let $X_1(k)$ and $X_2(k)$ be any two input signals, and let $Y_1(k)$ and $Y_2(k)$ be the respective output signals when these inputs are filtered. That is, $Y_1(k)$ and $Y_2(k)$ satisfy the following equations

$$Y_1(k) = X_1(k) + .5Y_1(k-1) \qquad (*)$$

$$Y_2(k) = X_2(k) + .5Y_2(k-1) \qquad (**)$$

Now consider the input signal

$$\widetilde{X}(k) = aX_1(k) + bX_2(k)$$

where a and b are any constants. We want to show that if $\widetilde{X}(k)$ is filtered, the output is

$$\widetilde{Y}(k) = aY_1(k) + bY_2(k)$$

This we can see by multiplying (*) by a, (**) by b, and adding equations, yielding

$$\widetilde{Y}(k) = \widetilde{X}(k) + .5\widetilde{Y}(k-1)$$

We can show that the filter is time-invariant in the same way. Let $X(k)$ and $Y(k)$ be the input and output of the filter, respectively. Then

$$Y(k) = X(k) + .5Y(k-1)$$

Now consider this equation with k replaced by $(k-r)$, where r is an integer:

$$Y(k-r) = X(k-r) + .5Y(k-r-1)$$

This last equation states that when $X(k-r)$ is applied as input to the filter, $Y(k-r)$ results as output, thus proving that the filter is time-invariant.

Exercise 5.1.1

Find the first 5 output sample values when the filter discussed above is used with the following input signals:

(a) $.5^k$

(b) k

(c) $\sin k\pi/4$

Exercise 5.1.2

Consider the digital filter whose defining equation is

$$Y(k) = X(k) + .8X(k-1)$$

(a) Find the transfer function as a function of z, where X is considered the input signal, and Y is the output signal.

(b) Find explicit expressions for the amplitude and phase response of this filter as functions of the frequency ω in radians/sample interval.

(c) Evaluate these expressions for 0 and π radians/sample interval.

Exercise 5.1.3

Repeat Exercise 5.1.2 for the filter whose defining equation is

$$Y(k) = X(k) - .8Y(k-1)$$

Discuss the relationship between the results in this and the preceding exercise.

5.2 RECURSIVE DIGITAL FILTERS OF GENERAL ORDER

The most general form of a recursive filter uses present and past values of the input, as well as past values of the output. We write this as

$$\begin{aligned}
Y(k) = &C(1)X(k) + \ldots + C(M)X(k-M+1) \\
&- D(1)Y(k-1) - \ldots - D(L)Y(k-L)
\end{aligned} \quad (5.2.1)$$

which has M arbitrary coefficients multiplying input values, and L arbitrary coefficients multiplying output values. If we take the z-transform of both sides of this equation, we can derive the transfer function as before. Thus

$$\begin{aligned}
Y^*(z) = &C(1)X^*(z) + \ldots + C(M)X^*(z)z^{-(M-1)} \\
&- D(1)Y^*(z)z^{-1} - \ldots - D(L)Y^*(z)z^{-L} \\
= &[C(1) + \ldots + C(M)z^{-(M-1)}]X^*(z) \\
&- [D(1)z^{-1} + \ldots + D(L)z^{-L}]Y^*(z)
\end{aligned} \quad (5.2.2)$$

Again, we now have an algebraic equation, which we can solve for the z-transform of the output:

$$Y^*(z) = \left[\frac{C(1) + \ldots + C(M)z^{-(M-1)}}{1 + D(1)z^{-1} + \ldots + D(L)z^{-L}} \right] X^*(z) \quad (5.2.3)$$

We see now that minus signs were chosen in (5.2.1) to define the coefficients $D(I)$ so that we would finally get plus signs in the rational transfer function.

Although M and L are arbitrary positive integers, it is convenient to take $M = L$ from now on. This assumption will make the discussion in terms of poles and zeros much simpler. Thus, the standard recursive digital filter will have the transfer function

$$H(z) = \frac{C(1) + \ldots + C(L)z^{-(L-1)}}{1 + D(1)z^{-1} + \ldots + D(L)z^{-L}} \qquad (5.2.4)$$

We shall call this form of recursive digital filter a *standard* recursive filter; that is, one with L denominator coefficients representing past output terms, and L numerator coefficients representing the present and $(L-1)$ past input terms, where $L \geq 1$. Such a filter is implemented in FORTRAN by the code

```
      DO 1 I=1,L
    1 Y(I)=0.
      LP=L+1
      DO 2 J=LP,N
      Y(J)=0.
      DO 2 I=1,L
      INDEX=J−I
    2 Y(J)=Y(J)+C(I)*X(INDEX+1)−D(I)*Y(INDEX)
```

$$(5.2.5)$$

The first L output values are taken to be zero as initial conditions. The output terms $Y(L+1)$ through $Y(N)$ are then computed using the filter equation (5.2.1) with $M = L$.

Let us now look at the transfer function of a standard recursive filter in terms of its zeros and poles. Multiplying numerator and denominator of (5.2.4) by z^{L}, we obtain

$$H(z) = \frac{z[C(1)z^{L-1} + \ldots + C(L)]}{z^{L} + D(1)z^{L-1} + \ldots + D(L)} \qquad (5.2.6)$$

Besides a factor of z in the numerator, corresponding to a zero at $z = 0$, we see that this is a ratio of an $(L-1)$st order polynomial and an Lth order polynomial. If we factor these polynomials, we can write

$$H(z) = \frac{zC(1) \cdot (z - zero(1))(z - zero(2)) \cdots (z - zero(L-1))}{(z - pole(1)) \cdots (z - pole(L))}$$

$$(5.2.7)$$

The factor $C(1)$ in the numerator arises from the fact that the leading term of the numerator is $C(1)z^{L-1}$. We shall assume in what follows that

$C(1) \neq 0$, and set $C(1) = const$. Hence, the standard recursive filter will have the transfer function

$$H(z) = \frac{const \cdot z \cdot (z - zero(1)) \cdots (z - zero(L-1))}{(z - pole(1)) \cdots (z - pole(L))} \qquad (5.2.8)$$

The amplitude and phase of the recursive filter transfer function can now be expressed in terms of the zeros and poles in a way similar to the moving average case. Thus, the phase of the transfer function is the sum of the phases of the numerator factors, minus the phases of the denominator factors. Taking $z = e^{j\omega}$, we can define the angles

$$\begin{aligned} \theta_i &= \text{Arg} \ (z - zero(i)) & i &= 1, \ldots, L-1 \\ \phi_i &= \text{Arg} \ (z - pole(i)) & i &= 1, \ldots, L \end{aligned} \qquad (5.2.9)$$

The interpretation of these angles is shown in Fig. 5.2.1.

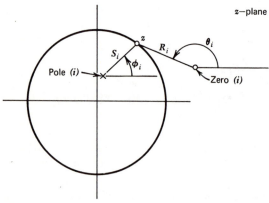

FIG. 5.2.1   The contributions of a pole and zero to the transfer function of a recursive filter: the factor (z-zero(*i*))/(z-pole(*i*)).

Hence

$$\begin{aligned} \text{Arg} \ H &= \text{Arg} \ (const) + \text{Arg} \ (z) + \theta_1 + \ldots + \theta_{L-1} \\ &\quad - \phi_1 - \ldots - \phi_L \\ &= \text{Arg} \ (const) + \omega + \theta_1 + \ldots + \theta_{L-1} \\ &\quad - \phi_1 - \ldots - \phi_L \end{aligned} \qquad (5.2.10)$$

where we have used the fact that $\text{Arg}(z) = \text{Arg}(e^{j\omega}) = \omega$. The Arg of *const* is, of course, $0°$ to $180°$, depending on whether *const* is positive or negative.

Similarly, the magnitude of $H(z)$ is the product of the magnitudes of the numerator factors, divided by the product of the denominator factors. These magnitudes are interpreted as the lengths of vectors from the zeros and the poles to the point z on the unit circle, as shown in Fig. 5.2.1. Thus

$$|H| = |const| \frac{R_1 \cdots R_{L-1}}{S_1 \cdots S_L} \qquad (5.2.11)$$

where we have used the fact that the magnitude of z is 1.

We see therefore that the presence of a zero close to the unit circle will cause the frequency response at the frequencies corresponding to the neighboring points on the unit circle to be small, while the presence of a pole near the unit circle will have the opposite effect. The effect of the zeros is the same as in the moving average filter, while the introduction of poles has a completely opposite effect, one that cannot be obtained with the use of zeros alone. This is the reason that recursive filters can have a wider variety of frequency response characteristics.

As an example, let us go back to the filter considered in Section 5.1:

$$H(z) = \frac{1}{(1 - .5z^{-1})} \qquad (5.2.12)$$

Multiplying numerator and denominator by z, this becomes in standard form:

$$H(z) = \frac{z}{(z - .5)} \qquad (5.2.13)$$

Thus, this filter has $L = 1$, $const = 1.$, and $pole$ $(1) = .5$ (see Fig. 5.2.2):

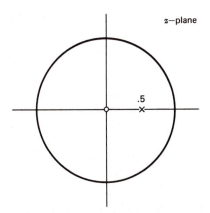

FIG. 5.2.2 Pole and zero of the transfer function of the recursive filter in the example.

This pole is nearer to the lower frequencies on the unit circle and, hence, will tend to accentuate these lower frequencies more than the higher ones. The zero at the origin will always be present in the standard form, and since it is equally distant from all points on the unit circle, it will not affect the magnitude of the transfer function.

Figure 5.2.3 shows a FORTRAN program that calculates the amplitude and phase of the transfer function of a standard recursive filter. This program is similar to the one for moving average filters, except that it reads in as data the zero and pole locations, rather than the coefficients. Figure 5.2.4 shows the amplitude and phase of the filter (5.2.13) used as an example above. The plot verifies our conclusion that the filter has a low pass effect. Since the pole is not very close to the unit circle, this effect is not pronounced.

```
C...AMPLITUDE AND PHASE OF THE DIGITAL FILTER WITH TRANSFER FUNCTION
C...H(Z) = CONST*Z*(Z-ZERO(1))...(Z-ZERO(L-1))/(Z-POLE(1))...(Z-POLE(L))
      COMPLEX ZERO(14),POLE(15),Z,G
      COMPLEX CEXP,CMPLX
      PI=3.141593
      CONVRT=180./PI
    1 READ(5,2)L,CONST
    2 FORMAT(I2,F10.5)
      WRITE(6,3)L,CONST
    3 FORMAT('1NUMBER OF POLES=',I3,' CONST=',E17.7)
      LM=L-1
      IF(L.EQ.1)GOTO6
      READ(5,4)(ZERO(I),I=1,LM)
    4 FORMAT(2F10.5)
      WRITE(6,5)(ZERO(I),I=1,LM)
    5 FORMAT(' ZEROS:'/(' ',E17.7,' +J ',E17.7))
    6 READ(5,4)(POLE(I),I=1,L)
      WRITE(6,7)(POLE(I),I=1,L)
    7 FORMAT(' POLES:'/(' ',E17.7,' +J ',E17.7))
      DO 12 J=1,101
      F=.01*FLOAT(J-1)
C......F IS FREQUENCY IN FRACTIONS OF THE NYQUIST FREQUENCY
      Z=CEXP(CMPLX(0.,F*PI))
      G=CMPLX(CONST,0.)*Z
      IF(L.EQ.1)GOTO9
      DO 8 I=1,LM
    8 G=G*(Z-ZERO(I))
    9 DO 10 I=1,L
   10 G=G/(Z-POLE(I))
      AMP=CABS(G)
      X=REAL(G)
      Y=AIMAG(G)
      IF(ABS(X).GT.1.E-8)GOTO11
      PHASE=SIGN(90.,Y)
      GOTO12
   11 PHASE=CONVRT*ATAN2(Y,X)
   12 WRITE(6,13)F,AMP,PHASE
   13 FORMAT(' F=',F5.2,' AMP=',E17.7,' PHASE=',E17.7)
      GOTO1
      END
```

FIG. 5.2.3 A FORTRAN program for calculating the amplitude and phase of a recursive filter.

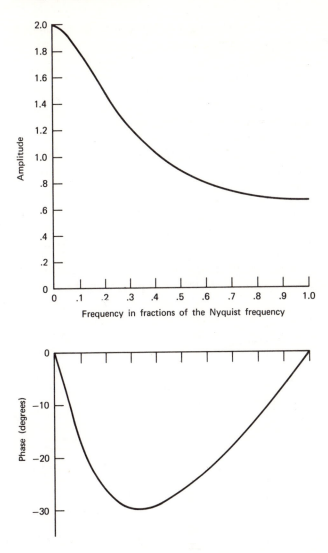

FIG. 5.2.4 The amplitude and phase of the recursive filter with transfer function given in Eq. 5.2.13. The pole and zero of this transfer function are shown in Fig. 5.2.2.

As another example of a recursive filter, we shall consider the standard recursive filter with the following parameters:

$$L = 5$$
$$const = .02486737$$
$$zero(1) = .9253846 + j.3790294$$
$$zero(2) = .9253846 - j.3790294$$
$$zero(3) = .6113717 + j.7913434$$
$$zero(4) = .6113717 - j.7913434 \qquad (5.2.14)$$
$$pole(1) = 0. + j0$$
$$pole(2) = .9312183 + j.2771898$$
$$pole(3) = .9312183 - j.2771898$$
$$pole(4) = .8645472 + j.1335386$$
$$pole(5) = .8645472 - j.1335386$$

The poles and zeros are plotted in the z-plane in Fig. 5.2.5. Notice that the pole at $z = 0$ cancels the zero at $z = 0$ that is present in the standard

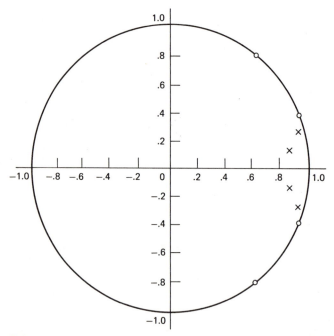

FIG. 5.2.5 Poles and zeros of the transfer function of the low pass recursive filter in the example, Eq. 5.2.14.

form, and there are altogether 4 poles and 4 zeros in the plane. These occur in complex conjugate pairs. Any complex poles or zeros will always occur in conjugate pairs if the filter equation is to have real coefficients. Notice also that the zeros occur precisely on the unit circle, and as shown in the plot of magnitude (Fig. 5.2.6), this makes the magnitude of the transfer function zero at these frequencies. This filter is the result of a rather elaborate design procedure, and is intended to pass frequencies from zero to one-tenth the Nyquist frequency, while suppressing all others.

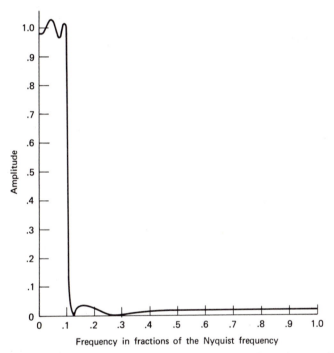

FIG. 5.2.6 The magnitude of the transfer function whose poles and zeros are shown in the previous figure.

Exercise 5.2.1

Can the rational transfer function (5.2.4) be derived using the phasor approach?

Exercise 5.2.2

Derive the transfer functions of the following filters and put them in the standard form of (5.2.4) if possible:

(a) $Y(k) = X(k-1) - .1Y(k-1)$
(b) $Y(k) = X(k-1) - .1Y(k-2)$
(c) $Y(k) = X(k) - .1Y(k-3)$
(d) $Y(k) = .5X(k) - 2Y(k-1) + Y(k-2)$
(e) $Y(k) = X(k) - X(k-1) - .5Y(k-1)$
(f) $Y(k) = X(k) - .5Y(k-1)$
(g) $Y(k) = X(k) + .2X(k-1) - .2Y(k-1)$

Exercise 5.2.3

Let the signal X defined by

$$X(0) = 1.$$
$$X(1) = -1.1$$
$$X(2) = .1$$
$$X(k) = 0. \qquad k > 2$$

be filtered by the recursive filter

$$Y(k) = X(k) + Y(k-1)$$

Find the output signal by the z-transform method, and check the result using the filtering equation. Can you explain the result in terms of zeros and poles?

Exercise 5.2.4

Find the equations of the magnitude and phase curves plotted in Fig. 5.2.4.

Exercise 5.2.5 (computer experiment)

We shall study the frequency response of the simple two-pole recursive filter:

$$Y(k) = AX(k) - BY(k-1) - CY(k-2) \qquad (5.2.15)$$

which has the transfer function

$$H(z) = \frac{A}{1 + Bz^{-1} + Cz^{-2}} \qquad (5.2.16)$$

We assume that the poles are complex, so that this transfer function can be written

$$H(z) = \frac{Az^2}{(z - Re^{j\theta})(z - Re^{-j\theta})} = \frac{Az^2}{z^2 - 2R \cos \theta z + R^2} \qquad (5.2.17)$$

This means that the poles occur in the complex z-plane at angles $\pm \theta$ and at a distance R from the origin (see Fig. 5.2.7).

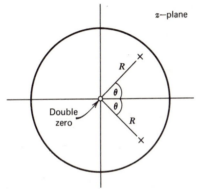

FIG. 5.2.7   Poles and zeros of the two-pole resonator of Exercise 5.2.5.

Such a filter is called a *resonator*, because it tends to have a large amplitude response at frequencies in the regions on the unit circle near the poles (i.e., to *resonate* at these frequencies).

(a) Choose various values of R and θ, and for the corresponding filters calculate the output signal corresponding to the following input:

$$X(0) = 1.$$
$$X(k) = 0. \qquad k \neq 0$$

(Such an input is called the *unit impulse*.) Verify the fact that the output decays to zero if $R < 1$, and explodes if $R > 1$. To calculate the output, implement the filter in FORTRAN.

(b) Write a simple FORTRAN program that calculates the amplitude of the frequency response of these filters. Correlate the value of R with the sharpness of the peak in the amplitude response, and with the rate at which the output signal decays or grows.

Exercise 5.2.6

A unit step input signal is applied to a 2-pole recursive filter with poles at $z = .5e^{j\pi/4}$ and $.5e^{-j\pi/4}$, and a numerator coefficient of unity. Find the first 4 terms of the output signal.

Exercise 5.2.7

Consider the two-pole digital filter with transfer function

$$H(z) = \frac{A}{1 + Bz^{-1} + Cz^{-2}}$$

and with poles at $z = Re^{\pm j\theta}$.

(a) Show that the filter coefficients B and C are related to R and θ by

$$B = -2R \cos \theta$$
$$C = R^2$$

(b) Show that the squared magnitude of the frequency response is given by

$$|H(e^{j\phi})|^2 = \frac{A^2}{(\cos 2\phi + B \cos \phi + C)^2 + (\sin 2\phi + B \sin \phi)^2}$$

where ϕ is the frequency in radians/sample interval.

*(c) It is sometimes convenient to choose A so that the maximum value of its amplitude response is about one. We shall do this by imposing the condition

$$|H(e^{j\theta})|^2 = 1$$

Show that this leads to a value

$$A = (1 - R) \sqrt{1 + R^2 - 2R \cos 2\theta}$$

*(d) Show that the bandwidth of the filter, defined to be the width of the amplitude response curve between points where the magnitude-square of the transfer function is equal to one-half its maximum, is approximately

$$2(1 - R) \qquad \text{radians/sample interval}$$

for R close to 1.

*(e) Show that the peak of the amplitude response characteristic occurs precisely at the frequency

$$\cos^{-1}\left(\frac{1+R^2}{2R}\cos\theta\right) \qquad \text{radians/sample interval}$$

Estimate the error in Hz involved in assuming the peak occurs at θ radians/ sample interval when the Nyquist frequency is 5000 Hz, $\theta = 500$ Hz, and the bandwidth as defined in part (d) is 200 Hz.

Exercise 5.2.8

A recursive digital filter is implemented by the following difference equation:

$$Y(k) = X(k) - AY(k-1) - BY(k-2)$$

where X and Y are the input and output signals, respectively.

(a) Find values of A and B such that the filter has poles on the imaginary axis in the z-plane, at a distance .9 from the origin.

(b) Calculate the amplitude and phase of the frequency response at the frequencies

$$\omega = 0 \qquad \pi/2, \qquad \text{and} \qquad \pi \text{ radians/sample interval}$$

(c) Sketch the amplitude of the frequency response from $\omega = 0$ to π radians/sample interval.

(d) If the sampling interval is 0.001 second, what does the frequency $\omega = \pi/2$ correspond to in Hz?

Exercise 5.2.9

Consider a recursive digital filter with transfer function in the standard form of (5.2.4) with L even. Show that we can write it in the factored form

$$G(z) = const \cdot \prod_{k=1}^{L/2}\left(\frac{1+a_k z^{-1}+b_k z^{-2}}{1+c_k z^{-1}+d_k z^{-2}}\right)$$

where the a_k, b_k, c_k, and d_k are real. This shows that we can implement such a standard recursive digital filter as a chain of filters

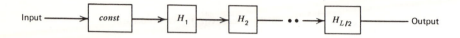

where each H_k has the transfer function

$$H_k(z) = \frac{1 + a_k z^{-1} + b_k z^{-2}}{1 + c_k z^{-1} + d_k z^{-2}}$$

Such an implementation is called the *cascade form* of the digital filter G, and has practical interest for a number of reasons, some of which we can just mention here. First, it turns out that the defining equation of a recursive filter (5.2.1), which is called the *direct form*, requires that we store the coefficients $C(I)$ and $D(I)$ to very high precision when the filter is of high order. This effect is, in general, lessened when the cascade form is used. Another consideration is the possibility of using special purpose, modular hardware to implement digital filters without using a general purpose computer.

Exercise 5.2.10

The trapezoidal rule for numerical integration is

$$y_k = \frac{(f_k + f_{k-1})}{2} + y_{k-1}$$

where f_k are values at equally spaced points of the function to be integrated, and y_k are values of the estimated integral. This operation can be viewed as a linear time-invariant digital filter H with input f and output y.

(a) Find the transfer function of this filter, $H(z)$.

(b) Derive a simple analytical expression for the magnitude of the transfer function as a function of the frequency ω. Sketch the magnitude characteristic for $0 \leq \omega \leq \pi$ radians/sample interval.

(c) Find and sketch the phase angle of the transfer function.

(d) Sketch as a function of ω the magnitude and phase of $1/j\omega$, the transfer function of a continuous-time integrator, and compare with (c).

5.3 INVERSE z-TRANSFORM, METHOD 2: PARTIAL FRACTION EXPANSION

We have seen in the compilation of our table of z-transforms, Fig. 4.4.10, that every signal of exponential or sinusoidal form has a z-transform that is a rational function of z. This holds even for arbitrary sums of such signals, since the sum of rational functions is also a rational function. Furthermore, every moving average or recursive filter is represented by multiplication by a rational function of z (we take the polynomial in z^{-1} that is the transfer function of a moving average filter to be a special case of a rational function). Hence, every situation that we shall encounter will result in a signal with a rational z-transform. We have already studied one method for taking the inverse z-transform of a rational function: the method of long division. Although this method provides us with numerical values for the sample values of the inverse transform, it does not give much insight into the composition of the signal. For example, in taking the inverse transform of the function of Section 5.1:

$$Y^*(z) = \frac{1}{(1 - z^{-1})(1 - .5z^{-1})} \tag{5.3.1}$$

we found that the signal $Y(k)$ seemed to approach the value 2 as a limit in an exponential fashion, but we did not obtain an analytical form that verified this. The second method of taking the inverse z-transform involves breaking the signal down into simple exponential components and, hence, gives us much valuable information about our answer.

To illustrate the method, consider the same rational transform mentioned above, (5.3.1). First, let us write this transform in standard form by multiplying numerator and denominator by z^2:

$$Y^*(z) = z \left[\frac{z}{(z-1)(z-.5)} \right] \tag{5.3.2}$$

The function inside the bracket has the degree of its numerator less than the degree of its denominator. Such rational functions are called *proper* rational functions, and are very important for our purposes. The reason they are important is that they can always be represented as a sum of simple one-pole terms as follows:

$$\frac{z}{(z-1)(z-.5)} = \frac{A}{(z-1)} + \frac{B}{(z-.5)} \tag{5.3.3}$$

This is *not* true for rational functions that are *not* proper, and this fact accounts for our need to put rational functions in standard form. We are left with the problem of solving for the constants A and B in (5.3.3). To do this, multiply the equation by the denominator of the rational function:

$$z = A(z - .5) + B(z - 1) \qquad (5.3.4)$$

If we now set $z = 1$ in this expression, the term involving B becomes zero, and we get one equation with the single unknown A:

$$1 = A(1 - .5) \qquad (5.3.5)$$

Hence,

$$A = 2 \qquad (5.3.6)$$

Similarly, if we set $z = .5$, we eliminate the term involving A and we can solve for B:

$$.5 = B(.5 - 1) \qquad (5.3.7)$$

so that

$$B = -1 \qquad (5.3.8)$$

Therefore (5.3.3) becomes

$$\frac{z}{(z-1)(z-.5)} = \frac{2}{(z-1)} - \frac{1}{(z-.5)} \qquad (5.3.9)$$

When this is substituted into (5.3.2) the original z-transform becomes

$$Y^*(z) = z\left[\frac{2}{(z-1)} - \frac{1}{(z-.5)}\right] = \frac{2}{1-z^{-1}} - \frac{1}{1-.5z^{-1}} \qquad (5.3.10)$$

Now each of these terms is in our table as the following z-transform:

$$c^k \xrightarrow{\quad Z \quad} \frac{1}{1-cz^{-1}} \qquad (5.3.11)$$

Hence the inverse z-transform of (5.3.11) can be written immediately as

$$Y(k) = 2 - (.5)^k \qquad (5.3.12)$$

This verifies our observation that the signal approaches the value 2 in an exponential fashion, and gives us an analytical expression for the signal $Y(k)$ for every k.

Now the procedure used above can be generalized as follows:

Partial Fraction Expansion Theorem

Consider the z-transform in standard form with L poles:

$$Y^*(z) = \frac{const \cdot z \cdot (z - zero(1)) \cdots (z - zero(L-1))}{(z - pole(1)) \cdots (z - pole(L))} \qquad (5.3.13)$$

Then *if the* L *poles are all different,* the inverse z-transform is

$$Y(k) = const \cdot [res(1) \cdot (pole(1))^k + res(2) \cdot (pole(2))^k + \cdots + res(L) \cdot (pole(L))^k] \quad (5.3.14)$$

where the coefficients $res(i)$, $i = 1, \ldots, L$ are called residues, and are calculated by the formula:

$$res(i) = \underbrace{\frac{(pole(i) - zero(1)) \cdots (pole(i) - zero(L-1))}{(pole(i) - pole(1)) \cdots (pole(i) - pole(L))}}_{\text{denominator does not contain the factor}} \quad (5.3.15)$$

$$(pole(i) - pole(i))$$

Proof

The proof follows the steps in the example above. Thus, the proper part of (5.3.13) is written

$$\frac{(z - zero(1)) \cdots (z - zero(L-1))}{(z - pole(1)) \cdots (z - pole(L))} = \frac{res(1)}{z - pole(1)} + \cdots + \frac{res(L)}{z - pole(L)} \quad (5.3.16)$$

Multiplying by the denominator and setting $z = pole(i)$ yields the formula (5.3.15). The original $Y^*(z)$ is then obtained by multiplying by $const \cdot z$, and this yields:

$$Y^*(z) = const \cdot \left(\frac{res(1)}{1 - z^{-1}pole(1)} + \cdots + \frac{res(L)}{1 - z^{-1}pole(L)} \right) \quad (5.3.17)$$

The transform (5.3.11) can then be applied term by term to yield the total inverse z-transform (5.3.14).

It is important to note the restriction that the poles all be different. The method can be extended to the case where multiple poles occur, but the algebra is more complicated. We shall be content to deal with the case of distinct poles only.

The procedure in the partial fraction expansion theorem is valid for complex poles. The inverse z-transform in this case has for each complex pair of poles two terms, each of which is the complex conjugate of the other. This ensures that the signal will be real. To illustrate the operation of partial fraction expansion with complex poles, let us calculate the inverse z-transform of the simple rational function:

$$Y^*(z) = \frac{1}{1 - z^{-1} + .5z^{-2}} \quad (5.3.18)$$

To apply the technique, we must factor the denominator:

$$Y^*(z) = \frac{z \cdot z}{(z - .5 - j.5)\ (z - .5 + j.5)} \tag{5.3.19}$$

Hence, in standard form this transform is described by the parameters

$$
\begin{aligned}
&L = 2 \\
&const = 1 \\
&zero(1) = 0 \\
&pole(1) = .5 + j.5 \\
&pole(2) = .5 - j.5
\end{aligned}
\tag{5.3.20}
$$

Writing the proper part as a sum of simple terms, we get

$$\frac{z}{(z - .5 - .5j)\ (z - .5 + j.5)} = \frac{A}{z - .5 - j.5} + \frac{B}{z - .5 + j.5} \tag{5.3.21}$$

Multiplying by the denominator, we get

$$z = A(z - .5 + j.5) + B(z - .5 - j.5) \tag{5.3.22}$$

Setting $z = .5 + j.5$, we can solve for A

$$A = \frac{.5 + j.5}{j} = .5 - j.5 = res(1) \tag{5.3.23}$$

Similarly, setting $z = .5 - j.5$ in (5.3.22), we obtain

$$B = \frac{.5 - j.5}{-j} = .5 + j.5 = res(2) \tag{5.3.24}$$

This verifies the fact that $res(2)$ is the complex conjugate of $res(1)$. The final inverse z-transform can then be written:

$$
\begin{aligned}
Y(k) &= res(1) \cdot (pole(1))^k + res(2) \cdot (pole(2))^k \\
&= (.5 - j.5)(.5 + j.5)^k + (.5 + j.5)(.5 - j.5)^k
\end{aligned}
\tag{5.3.25}
$$

Since these two terms are complex conjugates of each other, this can also be written as twice the real part of either:

$$Y(k) = 2\mathrm{Real}[\,(.5 - j.5)(.5 + j.5)^k\,] \tag{5.3.26}$$

If we wish to simplify this further, we can write

$$.5 + j.5 = .5(1 + j) = .5\sqrt{2}\left[\frac{1}{\sqrt{2}} + \frac{j}{\sqrt{2}}\right] = .5\sqrt{2}\,e^{j\pi/4} \tag{5.3.27}$$

and

$$
\begin{aligned}
Y(k) &= 2\mathrm{Real}[\,(.5 - j.5)(.5\sqrt{2})^k e^{jk\pi/4}\,] \\
&= \frac{1}{(\sqrt{2})^k}\left[\cos\frac{k\pi}{4} + \sin\frac{k\pi}{4}\right]
\end{aligned}
\tag{5.3.28}
$$

The first few terms of this can be checked by long division from the original z-transform (5.3.18).

Figure 5.3.1 shows a computer program that accepts a rational z-transform in terms of its poles and zeros, and calculates the partial fraction expansion and the inverse z-transform using the methods described above. The DO 9 loop calculates the numerator product of (5.3.15), while the DO 8 loop calculates the denominator. The residues are printed, and the inverse transform is calculated using (5.3.14) in the DO 13 loop. The residues include the factor CONST. The first 101 sample values are calculated, provided that the absolute value of the output signal does not exceed 1.E15, or drop below 1.E − 15 while different from zero. This test is done in the statement before the write statement 14. The imaginary part of the output, which should be precisely zero, is printed out as a

```
C...PARTIAL FRACTION EXPANSION AND INVERSE Z-TRANSFORM OF
C...H(Z) = CONST*Z*(Z-ZERO(1))...(Z-ZERO(L-1))/(Z-POLE(1))...(Z-POLE(L))
C...      = SUM OF RES(I)/(1-POLE(I)*Z**(-1))
      COMPLEX ZERO(14),POLE(15),RES(15),OUTPUT
    1 READ(5,2)L,CONST
    2 FORMAT(I2,F10.5)
      WRITE(6,3)L,CONST
    3 FORMAT('1NUMBER OF POLES=',I3,' CONST=',E17.7)
      IF(L.EQ.1)GOTO6
      LM=L-1
      READ(5,4)(ZERO(I),I=1,LM)
    4 FORMAT(2F10.5)
      WRITE(6,5)(ZERO(I),I=1,LM)
    5 FORMAT(' ZEROS:'/(' ',E17.7,' +J ',E17.7))
    6 READ(5,4)(POLE(I),I=1,L)
      WRITE(6,7)(POLE(I),I=1,L)
    7 FORMAT(' POLES:'/(' ',E17.7,' +J ',E17.7))
      DO 10 I=1,L
      RES(I)=CONST
      DO 8 J=1,L
      IF(J.EQ.I)GOTO8
      RES(I)=RES(I)/(POLE(I)-POLE(J))
    8 CONTINUE
      IF(L.EQ.1)GOTO10
      DO 9 J=1,LM
    9 RES(I)=RES(I)*(POLE(I)-ZERO(J))
   10 CONTINUE
      WRITE(6,11)(RES(I),I=1,L)
   11 FORMAT(' RESIDUES:'/(' ',E17.7,' +J ',E17.7))
      WRITE(6,12)
   12 FORMAT(' INVERSE Z-TRANSFORM:')
      DO 14 K=1,101
      N=K-1
      OUTPUT=0.
      DO 13 I=1,L
   13 OUTPUT=OUTPUT+RES(I)*POLE(I)**N
      Q=CABS(OUTPUT)
      IF(Q.GT.1.E15.OR.(Q.NE.0..AND.Q.LT.1.E-15))GOTO1
   14 WRITE(6,15)N,OUTPUT
   15 FORMAT(' N=',I3,' OUTPUT=',E20.10,' +J ',E20.10)
      GOTO1
      END
```

FIG. 5.3.1 A FORTRAN program for finding the partial fraction expansion and inverse z-transform of a rational z-transform, given its poles and zeros.

check. Usually, it is less than 1.E − 5, and this provides a monitor of the accumulation of round-off error.

To illustrate the operation of this program on a fairly complicated example, the output of the low pass filter discussed in Section 5.2 and illustrated in Figs. 5.2.5 and 5.2.6 was calculated when the input was the unit impulse: $X(k) = 1$ if $k = 0$, and 0 otherwise. The result is plotted in Fig. 5.3.2 and exhibits slow oscillatory behavior.

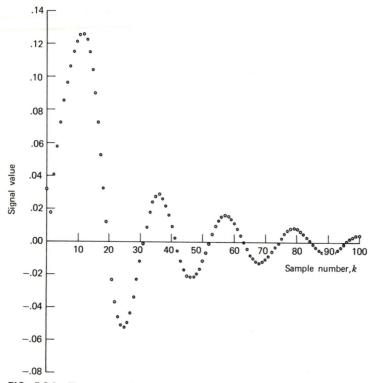

FIG. 5.3.2   The output signal of the low pass filter specified in Eq. 5.2.14 when the input signal is the unit impulse, calculated using the program shown in the previous figure. For this run we took *const*=.03395907.

Example Problem

Consider the digital filter with transfer function

$$\frac{H(z) = 1}{(1 + z^{-1})} \qquad (5.3.29)$$

The input signal is a unit step $X(k) = 1$, $k \geq 0$. Call the output signal $Y(k)$.

(a) Perform a partial fraction expansion of the output z-transform.

(b) Take the inverse z-transform of this partial fraction expansion term by term to obtain an expression for the kth sample value of Y.

(c) Check your answer to part (b) by using another method of taking the inverse z-transform.

Solution. The z-transform of a unit step is $1/(1 - z^{-1})$, so that the z-transform of the output signal is

$$Y^*(z) = H(z) X^*(z)$$

$$= \frac{1}{(1+z^{-1})(1-z^{-1})} = \frac{1}{1-z^{-2}} = \frac{z^2}{z^2-1}$$

$$= z\left[\frac{z}{z^2-1}\right] = z\left[\frac{A}{z-1} + \frac{B}{z+1}\right]$$

$$A = \frac{z}{z+1}\bigg]_{z=1} = 1/2 \tag{5.3.30}$$

$$B = \frac{z}{z-1}\bigg]_{z=-1} = 1/2$$

$$Y^*(z) = z\left[\frac{1/2}{z-1} + \frac{1/2}{z+1}\right] = \frac{1/2}{1-z^{-1}} + \frac{1/2}{1+z^{-1}}$$

We can take the inverse z-transform of this term by term, using Eq. 4.4.11 with $c = 1$ and -1. This gives

$$Y(k) = \tfrac{1}{2}[1 + (-1)^k] \tag{5.3.31}$$

By the method of long division, we obtain

$$
\begin{array}{r}
1+z^{-2}+z^{-4}+z^{-6}+z^{-8}+\cdots \\
1-z^{-2}\overline{\smash{\big)}\,1} \\
\underline{1-z^{-2}} \\
z^{-2} \\
\underline{z^{-2}-z^{-4}} \\
z^{-4} \\
\underline{z^{-4}-z^{-6}} \\
z^{-6} \\
\underline{z^{-6}-z^{-8}} \\
z^{-8}\cdots
\end{array}
\tag{5.3.32}
$$

so that the sample values of $Y(k)$ are

$$
\begin{array}{llllllllllll}
k & : & 0 & 1 & 2 & 3 & 4 & 5 & 6 & 7 & 8 & 9 \dots \\
Y(k): & & 1 & 0 & 1 & 0 & 1 & 0 & 1 & 0 & 1 & 0 \dots
\end{array}
$$

(5.3.33)

which agrees with Eq. 5.3.31.

Example Problem

A unit step digital signal $X(k) = 1$, $k \geq 0$, is applied to the digital filter H whose defining equation is

$$Y(k) = X(k) + 2Y(k-1), \qquad \text{where } Y \text{ is the output signal.}$$

(a) Find the z-transform $Y^*(z)$ of the output signal $Y(k)$.

(b) Find all the poles and zeros of $Y^*(z)$.

(c) Perform a partial fraction expansion of $Y^*(z)$ in terms of its poles.

(d) Take the inverse z-transform of $Y^*(z)$ term by term to obtain an expression for the kth sample value of the output signal $Y(k)$.

(e) State whether the filter H is stable or unstable.

Solution.

(a) The transfer function of the filter is

$$H(z) = \frac{1}{1 - 2z^{-1}}$$

(5.3.34)

so that

$$Y^*(z) = H(z) X^*(z) = \frac{1}{(1 - z^{-1})(1 - 2z^{-1})}$$

(5.3.35)

(b) Multiplying the numerator and denominator of $Y^*(z)$ by z^2, we obtain

$$Y^*(z) = \frac{z^2}{(z-1)(z-2)}$$

(5.3.36)

Hence there are two zeros at $z = 0$; and poles at $z = 1,2$.

(c) We want to find the coefficients A and B in

$$\frac{z}{(z-1)(z-2)} = \frac{A}{z-1} + \frac{B}{z-2}$$

(5.3.37)

From the partial fraction expansion theorem:

$$A = \left[\frac{z}{z-2}\right]_{z=1} = -1$$

$$B = \left[\frac{z}{z-1}\right]_{z=2} = 2$$

(5.3.38)

Therefore

$$Y^*(z) = z\left[\frac{z}{(z-1)(z-2)}\right] = z\left[\frac{-1}{z-1} + \frac{2}{z-2}\right] = \frac{-1}{1-z^{-1}} + \frac{2}{1-2z^{-1}}$$

(5.3.39)

(d) $Y(k) = -1 + 2 \cdot 2^k$ (5.3.40)

(e) H is unstable, since it has a pole at $z = 2$, outside the unit circle. Hence, $Y(k)$ grows without bound as k increases.

Exercise 5.3.1

Expand the following z-transforms in partial fractions, and write an expression for the kth term of the inverse z-transform:

(a) $\dfrac{1+z^{-1}}{1+z^{-2}}$

(c) $\dfrac{1}{1+z^{-3}}$

(b) $\dfrac{1}{(1-.5z^{-1})(1-.1z^{-1})}$

(d) $\dfrac{z^{-1}}{1+z^{-2}}$

Exercise 5.3.2

The signal

$$X(k) = \begin{cases} 0 & k<0 \\ 2^{-k} & k \geq 0 \end{cases}$$

is applied to a 2-pole recursive filter with poles at $z = 1, -1$, and a numerator coefficient of 1. Find an analytical expression for the output signal $Y(k)$ as a function of k.

Exercise 5.3.3

Consider the one-sided digital signal defined by

$$X(k) = \begin{cases} 1 & k=0, n, 2n, 3n, \ldots \\ 0 & \text{otherwise} \end{cases}$$

where n is a fixed positive integer.

(a) Find the z-transform of $X(k)$, $X^*(z)$, in closed form.

(b) For $n = 4$ find all the poles and zeros of $X^*(z)$ and plot their locations in the complex z-plane.

(c) For $n = 4$ sketch the magnitude of $X^*(z)$ for z on the unit circle (the frequency content), as a function of frequency ω from 0 to π radians/ sample interval.

Exercise 5.3.4

The first few Fibonacci numbers are:

$$1, 1, 2, 3, 5, 8, 13, 21, \ldots$$

Each is obtained by summing the two preceding numbers.

(a) Find the z-transform of the signal whose kth sample value is the kth Fibonacci number.

(b) Find the location of all the poles and zeros of this z-transform.

(c) Using a partial fraction expansion, derive an algebraic expression for the kth Fibonacci number.

Exercise 5.3.5

A baby has a bin of red blocks 1 unit high, a bin of green blocks 2 units high, and a bin of white blocks, also 2 units high. Find an analytic expression for the number of distinct piles he can build of height k. [*Hint:* if $N(k)$ is the number of distinct piles of height k, then $N(k) = N(k - 1) + 2N(k - 2)$.]

Exercise 5.3.6

Show that the transfer function of a standard recursive filter with real coefficients can be written in the form

$$G(z) = \sum_{k=1}^{L} \frac{a_k + b_k z^{-1}}{1 + c_k z^{-1} + d_k z^{-2}}$$

where a_k, b_k, c_k, and d_k are real. This shows that we can implement such a standard recursive digital filter by adding outputs of simple filters H_k as shown below:

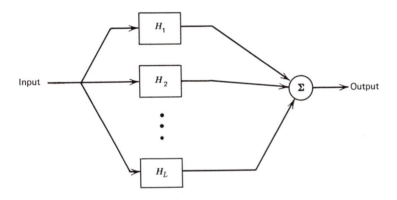

where each H_k has

$$H_k(z) = \frac{a_k + b_k z^{-1}}{1 + c_k z^{-1} + d_k z^{-2}}$$

Such an implementation is called the *parallel form* of the digital filter G, and has advantages similar to the cascade form (see Exercise 5.2.9).

Exercise 5.3.7

The signal $X(k) = a^k$, $k \geq 0$, is applied to the digital filter H whose defining equation is

$$Y(k) = X(k) + bY(k-1)$$

where Y is the output signal, $Y(-1) = 0$, and $a \neq b$.

(a) Find the z-transform $Y^*(z)$ of the output signal $Y(k)$.

(b) Find a partial fraction expansion of $Y^*(z)$ in terms of its poles.

(c) Find the inverse z-transform of $Y^*(z)$ by taking the inverse z-transform of its partial fraction expansion, yielding an algebraic expression for the kth sample value of the output signal $Y(k)$.

* (d) Find a simple expression for $Y(k)$ in the case $a = b$ by fixing b and taking the limit as $a \to b$ of the answer to part (c).

Exercise 5.3.8

The digital signal $X(k)$ has z-transform

$$X^*(z) = \frac{1}{1 + .25z^{-2}}$$

(a) Find the inverse z-transform $X(k)$ using a partial fraction expansion.

(b) Repeat using long division and compare the result.

Exercise 5.3.9

The recursive digital filter H has the defining equation

$$Y(k) = X(k) + .25\, Y(k-2)$$

(a) Find the transfer function $H^*(z)$.

(b) Find the location of all the poles and zeros of H, and state whether H is stable or not, based on your answer.

(c) Find the inverse transform $H(k)$ as an explicit algebraic function of k, using a partial fraction expansion.

(d) Check your answer to (c) using long division.

(e) Find the magnitude of the transfer function at the frequencies $\omega = 0$, $\pi/2$, π radians/sample interval; and sketch the magnitude of the transfer function for the frequency range 0 to π radians/sample interval.

*Exercise 5.3.10

Find an analytical expression for the signal $X(k)$ whose z-transform is

$$X^*(z) = \frac{1}{(1-z^{-1})^2(1+z^{-1})}$$

(*Hint:* This z-transform has a repeated pole and requires special treatment. Write

$$X^*(z) = z\left[\frac{z^2}{(z-1)^2(z+1)}\right]$$

and assume that the term in the brackets can be expanded in a partial fraction expansion of the following form:

$$\frac{z^2}{(z-1)^2(z+1)} = \frac{A}{z-1} + \frac{B}{(z-1)^2} + \frac{C}{(z+1)}$$

Check your answer for the first few sample numbers using long division. This approach can be used to generalize the method given in the text to transforms with repeated poles.)

5.4 STABILITY

We now take up the important question of whether a particular signal increases in magnitude without bound, or decays to zero. Consider any signal X whose z-transform is a rational function of z with L poles. (We shall assume from now on that these L poles are all different.) From the partial fraction expansion theorem we know that $X(k)$ is the sum of L terms of the form

$$const \cdot (pole)^k \qquad\qquad (5.4.1)$$

Now for each such term there are three possibilities:

(1) $|pole| < 1$. In this case the term associated with this pole decreases in magnitude to zero as k increases.

(2) $|pole| = 1$. Here the term associated with this pole remains the same in magnitude as k increases. When two such terms are complex conjugates of one another, the sum combines to form a term of the form $const \cdot \sin(k\omega + \theta)$.

(3) $|pole| > 1$. In this case the term associated with this pole increases in magnitude without bound as k increases.

We see then that the behavior of the signal as k increases is determined by the pole with the largest magnitude. We then have three cases, depending on the pole with the largest magnitude:

(1) The pole with the largest magnitude is inside the unit circle. In this case we say the signal is *stable*. It decreases to zero.

(2) The pole with the largest magnitude is on the unit circle. Here we say that the signal is *marginally stable*. It remains bounded as k increases.

(3) The pole with the largest magnitude is outside the unit circle. In this case we say that the signal is *unstable*. It diverges to infinity as k increases.

Thus, the pole positions of the z-transform of a signal tell us how the signal behaves as the sample number increases.

Now consider the case where a signal X with a rational transform is the input to a recursive filter. The output signal Y then has poles at the poles of the input signal X, and at the poles of the transfer function of the filter. Thus, the stability of the output signal depends both on the stability of the input signal, and on the pole locations of the transfer function of the filter. We introduce the concept of the stability of a *filter*:

Definition

A recursive filter is said to be *stable* if all the poles of its transfer function lie inside the unit circle in the z-plane.

This allows us to state that the output signal of a recursive filter is stable *if and only if* the input is stable, *and* the filter is stable.

Since we shall usually want to deal with stable signals (since unstable signals will cause overflow) we shall usually restrict our attention to recursive filters that are themselves stable. Thus, we shall want the poles of the transfer functions of any recursive filters to lie inside the unit circle. We see also from the fact that terms will be present in the output signal of the form $(pole)^k$, that the closer the poles of the filter are to the unit circle, the more slowly will the output signal decay to zero.

We should point out that when we say that the output signal is stable if and only if the filter and the input signal are stable, we assume that no cancellation takes place between a zero and a pole. More specifically, suppose the input signal transform has a zero at $z = 2$, and that the filter transfer function has a pole also at $z = 2$. Then upon multiplication the zero will cancel the pole, and there will not be a term in the output signal corresponding to the transfer function pole at $z = 2$. Thus, our statement is not strictly true, but it is practically true in the sense that *exact* cancellation is almost impossible. It requires that the zero and the pole agree to all significant figures, with the result that the term $(pole)^k$ has a coefficient exactly zero. We shall exclude this possibility in what we do.

We conclude with a comment about the problem of stability in more general systems. In time-invariant linear systems, as we have seen, the stability question is settled in an extremely simple way, merely by observing the positions of the poles of the z-transform of the signal and the transfer function involved. If the filter were nonlinear or time-varying, the problem of stability would become much more complicated, and a great deal of effort has been expended in studying the more general question of stability in nonlinear or time-varying systems.

Exercise 5.4.1

Show that a moving average filter is always stable in the sense that a stable input signal will always produce a stable output signal.

Exercise 5.4.2

Show that if a zero is present at the same position as a pole in a rational transform, that the residue corresponding to this pole will be zero.

Exercise 5.4.3

Show that a signal can have a zero outside the unit circle and still decay to zero.

Exercise 5.4.4

The recursive digital filter H has the transfer function

$$H(z) = \frac{a_0 + a_1 z^{-1} + \ldots + a_{n-1} z^{-(n-1)}}{1 + b_1 z^{-1} + \ldots + b_n z^{-n}}$$

H has all its poles strictly inside the unit circle. The unit step signal $X(k) = 0$, $k < 0$; $X(k) = 1$, $k \geq 0$; is applied as an input. Show that if $Y(k)$ is the output signal, then

$$\lim_{k \to \infty} Y(k) = H(1)$$

(*Hint:* Use a partial fraction expansion.)

Exercise 5.4.5 (computer experiment)

Perform the experiment of filtering a signal (any signal) with a recursive filter you know to be unstable. Predict the results beforehand.

5.5 IMPULSE RESPONSE AND CONVOLUTION

Consider now the transfer function of a digital filter, $H(z)$. The inverse z-transform of $H(z)$ is the output signal when the input signal is the unit impulse signal defined as follows:

$$X(k) = \begin{cases} 1 & k = 0 \\ 0 & \text{otherwise} \end{cases} \tag{5.5.1}$$

since in this case $X^*(z) = 1$. This output signal, termed the *impulse response* of the filter H, has a special significance: It determines the response of the filter to an arbitrary signal.

To see this, denote the impulse response of the filter H by $H(k)$. Now let $X(k)$ be any input signal whatsoever. Consider first the response of the filter to the first sample value $X(0)$. Since the response to an input signal at $k = 0$ of unit magnitude is $H(k)$, the response to a signal of size $X(0)$ at $k = 0$ will be $X(0)H(k)$. Now consider the response of the filter to the next input sample value, $X(1)$. Since the filter is time-invariant, the response will be a shifted version of the response of the filter to such a sample value at $k = 0$. Specifically, the response to $X(1)$ at $k = 1$ will be $X(1)H(k - 1)$. In general, the response of the filter to the input sample value $X(j)$ will be $X(j)H(k - j)$. We now think of the input signal as being composed of a sum of signals, each zero everywhere except at the sample number j. The output signal will then be a sum of the corresponding output terms, by the linearity of the filter. Thus the output signal will be

$$Y(k) = X(0)H(k) + X(1)H(k-1) + X(2)H(k-2) + \ldots$$
$$= \sum_{j=0}^{\infty} X(j)H(k-j) \tag{5.5.2}$$

This important formula is called a *convolution* between the signals X and H. Notice that it involves the sum of terms $X(j)H(k-j)$, where k is the output sample number under consideration. The index of summation appears with a negative sign in H, the impulse response, and this means that the plot of the impulse response is reversed and then multiplied term by term with the input.

To illustrate, consider the case where $X(k)$ is the unit step and $H(k) = (.5)^k$. This corresponds to the transforms

$$X^*(z) = \frac{1}{1-z^{-1}}$$
$$H(z) = \frac{1}{1-.5z^{-1}} \tag{5.5.3}$$

The convolution formula (5.5.2) will now be evaluated in the case $k = 4$, to find the output signal $Y(4)$. The plots in Fig. 5.5.1 show $X(j)$, $H(j)$, and $H(4-j)$.

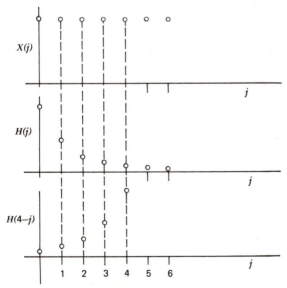

FIG. 5.5.1 Illustration of the convolution of the signal $X(k)$ with the impulse response of the filter H.

If we now multiply $X(j)$ by $H(4-j)$ term by term, we obtain the following terms:

$$1, .5, .25, .125, .0625 \qquad (5.5.4)$$

Adding these up we get $Y(4) = 1.9375$, which checks (5.1.9).

Since the plot of H is reversed in time, the convolution formula is sometimes called the *Faltung* of the signals X and H, from the German word for folding. In fact, one meaning of the word "convolute" is "rolled or wound together one part upon another" (*Webster's Seventh New Collegiate Dictionary*, G. & C. Merriam Company, 1967). This is precisely what the convolution formula expresses.

It should be pointed out that the summation in the convolution formula is not really infinite in a practical situation, since $H(k)$ vanishes for negative k. This is a consequence of the fact that the filter H cannot respond to an input before it occurs, so that the impulse response is one-sided just as are the other digital signals we deal with. Hence the upper limit can be changed from infinity to k; after this point $H(k-j)$ is zero.

Another important observation concerns the symmetry of the situation. Suppose the input signal had a z-transform $H(z)$ and the filter transfer function were $X^*(z)$. Then the output z-transform would still be $X^*(z)H(z)$ and, therefore, the output signal would be the same. Hence we may reverse the roles of X and H in the convolution formula, obtaining:

$$Y(k) = \sum_{j=0}^{\infty} X(j) H(k-j) = \sum_{j=0}^{\infty} H(j) X(k-j) \qquad (5.5.5)$$

Because of this symmetry, we can represent the convolution of the two signals $H(k)$ and $X(k)$ by the expression

$$H(k) \otimes X(k) = X(k) \otimes H(k) \qquad (5.5.6)$$

Since the z-transform of $Y(k)$ is the product of the z-transforms of X and H, we have

$$X(k) \otimes H(k) \xrightarrow{\quad Z \quad} X^*(z) H(z) \qquad (5.5.7)$$

We have therefore shown that the z-transform of a convolution is the *product* of the respective z-transforms. Thus, the z-transform can be interpreted as a transformation that converts convolution to multiplication.

The second form of the convolution in (5.5.5) shows still another interpretation. This is exactly the form of a moving average filter with coefficients equal to the impulse response $H(k)$ of the filter. Hence, a recursive filter can be thought of as a moving average filter with an infinite number of coefficients. This is consistent with the operation of expanding the transfer function in an infinite series. Thus, if $H(z)$ is taken as above to be $1/(1 - .5z^{-1})$ we have

$$\frac{1}{1-.5z^{-1}} = 1 + .5z^{-1} + .25z^{-2} + \ldots \tag{5.5.8}$$

and

$$Y(k) = \sum_{j=0}^{\infty} (.5)^j X(k-j) \tag{5.5.9}$$

Hence, the term $H(j)$ in the impulse response of a recursive filter represents the weight given to an input term j sampling intervals in the past.

Exercise 5.5.1

Take the z-transform of the convolution formula (either version of Eq. 5.5.5) and from this show that $Y(z) = X^*(z)H(z)$. This involves the use of Property 1 of the z-transform.

Exercise 5.5.2

Show that the derivation of Eq. 5.5.2 made use of the fact that the filter was linear and time-invariant, as described in Section 3.1. Does the derivation work for any linear time-invariant filter?

Exercise 5.5.3

Let $X(k)$ and $Y(k)$ be the input and output signals, respectively, of the digital filter with transfer function $H(z) = 1/(1 - z^{-1})$. Show that

$$Y(k) = \sum_{i=0}^{k} X(i)$$

Exercise 5.5.4

Theoretically, if we pass a signal through the filter with transfer function $H(z)$, and then pass the result of that operation through the filter with transfer function $1/H(z)$, we should get back to our original signal. In some cases this will work well, and in others not. Discuss this problem, giving conditions that ensure it working well.

Exercise 5.5.5

Give an example of a recursive filter that has a finite-duration impulse response. That is, for some M, $H(k) = 0$ for all $k \geq M$.

6.
FREQUENCY ANALYSIS

6.1 THE DISCRETE FOURIER TRANSFORM (DFT)

Suppose we are given N sample values of a digital signal, say

$$F(0), F(1), \ldots, F(N-1) \tag{6.1.1}$$

and we wish to calculate its z-transform numerically. That is, we wish to calculate the z-transform

$$F^*(z) = \sum_{n=0}^{N-1} F(n) z^{-n} \tag{6.1.2}$$

for z at various points on the unit circle. The magnitude of this will be the frequency content, or *spectrum,* of the given segment of the signal. Such analysis is useful in many fields of study where the frequency content of a signal provides information that enables us to draw conclusions about the origin of the signal or to characterize the signal in a convenient way. For example, the spectrum of oceanographic waves may enable us to isolate the frequency components caused by the moon tide, sun tide, earthquakes, and random effects. As another example, the frequency analysis of speech waveforms enables us to characterize different vowels by certain peaks in their spectra (called *formants*).

Notice from (6.1.2) that we have assumed that the signal values are zero outside the range 0 to $N-1$. In this sense the transform is a *finite transform*; it deals only with a finite record of the signal.

Now we cannot calculate the value of the transform (6.1.2) at all frequencies ω, but only at some finite set. We choose the set of N frequencies

$$\omega_k = \frac{k2\pi}{N} \qquad k = 0, \ldots, N-1 \tag{6.1.3}$$

which represent points equally spaced around the unit circle, as shown in Fig. 6.1.1 for $N = 16$:

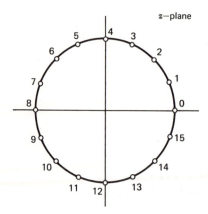

FIG. 6.1.1 The set of points on the unit circle in the z-plane where the Discrete Fourier Transform is calculated.

These correspond to values of z given by

$$z = e^{jk2\pi/N} \qquad k = 0, \ldots, N-1 \qquad (6.1.4)$$

and the transform

$$F^*(k) = \sum_{n=0}^{N-1} F(n) e^{-jnk2\pi/N} \qquad k = 0, \ldots, N-1 \qquad (6.1.5)$$

We have used the symbol $F^*(k)$ to represent $F^*(z = e^{jk2\pi/N})$. One reason for choosing to calculate F^* at N equally spaced frequency points is that the inverse transform can be calculated very simply. That is, given the N values of the transform $F^*(0), \ldots, F^*(N-1)$, we can calculate the original N values of the signal. The process of calculating the N values of the transform is called the *Discrete Fourier Transform* (DFT) and the inverse operation is called, naturally enough, the *Inverse Discrete Fourier Transform* (IDFT). As mentioned above, the IDFT has a simple form, which we express as the following theorem.

Theorem

Given N points of the DFT

$$F^*(k) = \sum_{n=0}^{N-1} F(n) e^{-jnk2\pi/N} \qquad k = 0, \ldots, N-1 \qquad (6.1.6)$$

the IDFT is given by

$$F(p) = \frac{1}{N} \sum_{k=0}^{N-1} F^*(k) e^{jkp2\pi/N} \qquad p = 0, \ldots, N-1 \qquad (6.1.7)$$

(Notice that these formulas are identical in form, except for a factor of $1/N$ and the sign in the exponent.)

Proof

Multiply (6.1.6) by $e^{jkp2\pi/N}$ for $p = 0, \ldots, N-1$, and sum from $k = 0$ to $N - 1$, yielding

$$\sum_{k=0}^{N-1} F^*(k) e^{jkp2\pi/N} = \sum_{k=0}^{N-1} \sum_{n=0}^{N-1} F(n) e^{jk(p-n)2\pi/N} \qquad p = 0, \ldots, N-1$$
$$(6.1.8)$$

If we reverse the order of summation, the right-hand side becomes

$$\sum_{n=0}^{N-1} F(n) \left[\sum_{k=0}^{N-1} e^{jk(p-n)2\pi/N} \right] \qquad p = 0, \ldots, N-1 \qquad (6.1.9)$$

We now investigate the bracketed summation

$$\sum_{k=0}^{N-1} e^{jk(p-n)2\pi/N} \qquad p, n = 0, \ldots, N-1 \qquad (6.1.10)$$

where we must consider all values of both p and n from 0 to $N - 1$. If $p = n$, we have simply

$$\sum_{k=0}^{N-1} 1 = N \qquad \text{if} \qquad p = n \qquad (6.1.11)$$

If $p \neq n$, write (6.1.10) as

$$\sum_{k=0}^{N-1} a^k \qquad (6.1.12)$$

where

$$a = e^{j(p-n)2\pi/N} \qquad (6.1.13)$$

Since $p - n$ is between $-(N-1)$ and $+(N-1)$, $a \neq 1$. The partial geometric series (6.1.12) is a summation known from algebra (see Section 1.3),

$$\sum_{k=0}^{N-1} a^k = \frac{1 - a^N}{1 - a} \qquad a \neq 1 \qquad (6.1.14)$$

But the numerator vanishes, since

$$a^N = e^{j(p-n)2\pi} = 1 \tag{6.1.15}$$

Hence the bracketed summation (6.1.10) is N if $p = n$, and 0 otherwise. Eq. 6.1.8 can therefore be written

$$\sum_{k=0}^{N-1} F^*(k) e^{jkp2\pi/N} = NF(p) \tag{6.1.16}$$

which is what we wished to show.

The formula (6.1.7) for the IDFT strengthens our interpretation of the z-transform on the unit circle as the frequency analysis of a signal. It says that the signal $F(p)$ is the *sum of phasors of frequency $k2\pi/N$, each weighted by the factor $F^*(k)$.*

The importance of the DFT has greatly increased in the last few years because of a very efficient way of calculating it. At first glance it seems that we need to perform N^2 multiplications to calculate (6.1.6): N multiplications for each of the N frequency points. It turns out, however, that if N is a power of 2, we need only $N \log_2 N$ multiplications. This efficient computational procedure is called the *Fast Fourier Transform* (FFT) and has made the DFT practical to use in many applications where it would otherwise be impractical. For example, if $N = 1024 = 2^{10}$, we need only $N \log_2 N = 10,240$ multiplications, instead of $N^2 = 1,048,576$, a savings of a hundredfold. The FFT algorithm will be derived in the next section.

Example

The FFT is implemented easily with small laboratory computers, so-called "minicomputers." Figure 6.1.2 shows the results of some analog-to-digital conversion experiments and FFT calculations performed at Princeton University with a Hewlett-Packard 2114B computer and an analog-to-digital converter.

Figure 6.1.2a shows the samples obtained from a sine wave of frequency about 1000 Hz. The sampling rate of the converter for all these experiments is 30,234 samples/second, corresponding to a Nyquist rate of 15,117 Hz. The converter has an accuracy of 12 bits including sign. Figure 6.1.2b shows the magnitude of the DFT versus frequency, computed using a 512-point FFT. This calculation took about 20 seconds, but could be

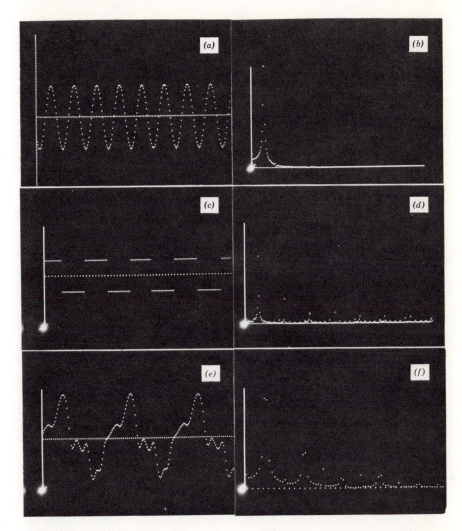

FIG. 6.1.2   Some results of analog-to-digital conversion and FFT calculation on a Hewlett-Packard 2114B minicomputer. The sampling frequency was 30,234 Hz. (*a*) Sinusoidal digital signal with frequency of about 1000 Hz; (*b*) Magnitude of the DFT of the previous signal, calculated using a 512-point FFT algorithm; (*c*) Square-wave digital signal with a repetition rate of about 1000 Hz; (*d*) Magnitude of the DFT of the previous signal; (*e*) Signal resulting from sampling a 440 Hz note played on a B-flat clarinet by Stephen Bradley; (*f*) Magnitude of the DFT of the previous signal.

speeded up considerably if fixed-point arithmetic were used instead of floating-point. Notice that the frequency content peaks sharply at the frequency of the original sine wave, as we would expect, and has no other significant peaks elsewhere.

Figure 6.1.2*c* and *d* show samples and frequency content, respectively, of a square wave of the same frequency as the sine wave above, about 1000 Hz. Notice now that the frequency content has significant peaks at the fundamental frequency of the square wave and at integral multiples of this frequency: the harmonics or "overtones" of the fundamental frequency. Furthermore, there is significant frequency content only at the odd-numbered harmonics, a property of square waves.

Figure 6.1.2*e* and *f* show samples and frequency content, respectively, of a 440 Hz note (concert A) played on a B-flat clarinet (courtesy of Stephen Bradley). The second harmonic seems to be absent, but the third to ninth are all discernible.

Exercise 6.1.1

Find the DFT of a record of N consecutive samples of a phasor at a frequency of $k2\pi/N$, where k is an integer.

Exercise 6.1.2

Repeat the previous exercise for a phasor at a frequency of $x2\pi/N$, where x is not necessarily an integer. Plot the magnitude of the DFT for a simple example of this case.

Exercise 6.1.3

(This exercise assumes familiarity with matrices.) The DFT can be considered as a multiplication of a signal vector by the matrix **F** whose entry in row i and column k is

$$e^{-j(ik)2\pi/N} \qquad i, k = 0, 1, \ldots, N-1$$

We shall number the rows and columns from 0 to $N - 1$.

(a) Show that $F^2 = NR$, where R is the matrix whose entry in row i and column k is 1 if $(i + k) = 0 \bmod N$, and 0 otherwise. That is,

$$R = \begin{bmatrix} 1\,0\,0\,0\,0\,...\,0 \\ 1 \\ 1 \\ 1 \\ ... \\ 0\,1\,0\,0\,0\,...\,0 \end{bmatrix}$$

(b) Show that $R^2 = I$, the identity matrix and, hence, that $F^4 = N^2 R^2 = N^2 I$.

(c) From $F^2 = NR$ show that $F^{-1} = (1/N)\,FR$.

(d) Show that the result of (c) is equivalent to (6.1.7).

*Exercise 6.1.4

Let $F(k)$ be any digital signal (possibly of infinite duration) and $F^*(z)$ its z-transform. Show that

$$F(k) = \frac{1}{2\pi} \int_{-\pi}^{\pi} F^*(e^{j\omega}) e^{jk\omega} d\omega$$

(*Hint:* this formula is analogous to (6.1.7) and can be derived in a similar way.) What interpretation can you give this formula?

Exercise 6.1.5 (computer experiment)

For small stretches of signal, say $N = 64$ or 128, it is practical to compute the DFT and its inverse directly from (6.1.6) and (6.1.7). Write a FORTRAN program that does this and run the following tests:
(*Hint:* For a given N, the complex exponentials need not be recomputed for each transform.)

(a) Verify the DFT for a constant and some phasors with different frequencies.

(b) Verify that the IDFT does in fact restore the original signal.

(c) Calculate the following sequence:
$A_1 = $ DFT of $F(k)$
$A_2 = $ IDFT of A_1
$A_3 = $ DFT of A_2

.

.

.

$A_r = $ IDFT of A_{r-1}

for as large an r as you can afford, and compare the result with the starting signal. The difference represents accumulated roundoff error.

(d) Calculate the DFT of the DFT of a signal. What is the effect of this operation in general (see Exercise 6.1.3)?

Exercise 6.1.6

(a) Assume that the finite length digital signal $F(n)$ is even in the sense that

$$F[\,(-n)\,(\mathrm{mod}\ N)\,] = F(n) \qquad n = 0, 1, \ldots, N-1$$

Show that this implies that its DFT, $F^*(k)$, is even in the same sense.

(b) Show that if $F(n)$ is real, $F^*(k)$ has an even real part and an odd imaginary part. Odd here means that $F[\,(-n)\,(\mathrm{mod}\ N)\,] = -F(n)$.

(c) Show that if $F(n)$ is both even and real, $F^*(k)$ is also even and real.

6.2 THE FAST FOURIER TRANSFORM

Consider the signal of Length N:

$$F(0), F(1), \ldots, F(N-1) \tag{6.2.1}$$

whose DFT we wish to compute. Let us split this signal into two signals as follows: let $G(n)$ be equal to the values of F at even sample numbers (assuming N is even):

$$G(n) = F(2n) \qquad n = 0, 1, \ldots, \frac{N}{2} - 1 \tag{6.2.2}$$

and let $H(n)$ be equal to the values of F at odd sample numbers:

$$H(n) = F(2n+1) \qquad n = 0, 1, \ldots, \frac{N}{2} - 1 \qquad (6.2.3)$$

Suppose now that we calculated the DFT of the signal $G(n)$, say $G^*(k)$. This is an $N/2$-point transform given by

$$G^*(k) = \sum_{n=0}^{N/2-1} G(n) e^{-jnk2\pi/(N/2)} \qquad k = 0, \ldots, \frac{N}{2} - 1 \qquad (6.2.4)$$

Notice that $(N/2)$ appears in this formula wherever we would ordinarily expect N, since this signal is of length $(N/2)$.

Now notice that if k is replaced by $(k + N/2)$, the expression (6.2.4) for $G^*(k)$ remains unchanged, since a multiple of $2\pi j$ is added to the exponent. Hence, we may consider $G^*(k)$ defined for $k = N/2, N/2 + 1, \ldots, N - 1$ by the relation

$$G^*(k + \frac{N}{2}) = G^*(k) \qquad (6.2.5)$$

We can compute $H^*(k)$ in the same way:

$$H^*(k) = \sum_{n=0}^{N/2-1} H(n) e^{-jnk2\pi/(N/2)} \qquad k = 0, 1, \ldots, \frac{N}{2} - 1 \qquad (6.2.6)$$

and again, this can be extended to the range $k = N/2, N/2 + 1, \ldots, N - 1$ by the periodicity relation

$$H^*\left(k + \frac{N}{2}\right) = H^*(k) \qquad (6.2.7)$$

Now (6.2.4) required $(N/2)^2$ multiplications, as did (6.2.6), so altogether we need $2(N/2)^2$ multiplications to obtain

$$\begin{array}{ll} G^*(k) & k = 0, \ldots, N-1 \\ H^*(k) & k = 0, \ldots, N-1 \end{array} \qquad (6.2.8)$$

We shall now show that G^* and H^* can be combined in a simple way to produce F^*. In particular, examine the combination

$$G^*(k) + e^{-jk2\pi/N} H^*(k) = \sum_{n=0}^{N/2-1} G(n) e^{-jnk2\pi/(N/2)} + $$
$$e^{-jk2\pi/N} \sum_{n=0}^{N/2-1} H(n) e^{-jnk2\pi/(N/2)} \qquad k = 0, 1, \ldots, N-1$$
$$(6.2.9)$$

Replacing $G(n)$ by its definition, $F(2n)$; $H(n)$ by $F(2n+1)$; and moving the exponential factor inside the right summation, we get

$$\sum_{n=0}^{N/2-1} F(2n)\, e^{-j(2n)k2\pi/N} + \sum_{n=0}^{N/2-1} F(2n+1)\, e^{-j(2n+1)k2\pi/N} \qquad (6.2.10)$$

The left-hand summation consists simply of the even-numbered terms in the DFT of F, while the right-hand summation consists of the odd-numbered terms. Thus

$$F^*(k) = G^*(k) + e^{-jk2\pi/N} H^*(k) \qquad (6.2.11)$$

which shows how the two "subtransforms," G^* and H^* can be combined to give the transform F^* of our original sequence. This process of calculating one N-point transform by combining two $N/2$-point transforms is called "merging." The merging process requires an additional N multiplications, so altogether, we have used $2(N/2)^2 + N$ multiplications. The most straightforward method of calculating an N-point transform, using the definition directly, requires N^2 multiplications, so we have saved $N^2 - (N^2/2 + N) = N^2/2 - N$ multiplications. Thus, when $N = 1000$, we save 499,000 multiplications (about half the total)!

Nothing stops us from computing each of the subtransforms G^* and H^* using the same trick. That is, divide $G(n)$ into two parts, even and odd; compute the DFT of each subsequence, and merge to get G^*. Then G^* requires $[2(N/4)^2 + N/2]$ multiplications, as does H^*, so the total for F^* becomes

$$4\left(\frac{N}{4}\right)^2 + 2N \qquad (6.2.12)$$

If N is a power of 2, we may continue this process $\log_2 N$ times. At the last step, the DFT for $N = 1$ requires no multiplications, and the N^2 term in the expression for the number of multiplications disappears. Each time we divide the signal into two parts, we introduce another N multiplications, and since we do this $\log_2 N$ times, we require altogether $N \log_2 N$ multiplications.

This algorithm is one form of the FFT (the so-called "decimation in time" algorithm). As mentioned above, it decreases the number of multiplications required by a factor of $N/\log_2 N$, which is a factor of more than a hundred for the reasonable-sized transform of size $2^{10} = 1024$. We shall assume in what follows that N is an integer power of 2, although similar savings are possible for any N.

Exercise 6.2.1

From the conditions $Q(n) = 2Q(n/2) + n$ and $Q(1) = 0$, prove that $Q(n) = n\log_2 n$, n an integer power of 2 greater than 1.

6.3 PROGRAMMING THE FFT

A wrinkle appears when we try to program this algorithm, which we must now deal with. Let us illustrate the problem with an 8-point transform. We need to calculate the DFT of the subsequence corresponding to the even-numbered points of our original F; and the DFT corresponding to the odd-numbered points. Therefore, we select the even-numbered points and put them in the first half of our array (producing G), and put the odd-numbered points in the second half of the array (producing H). This process is illustrated in Fig. 6.3.1. (We use subscripts in this section to simplify the notation.)

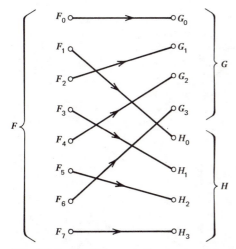

FIG. 6.3.1 Shuffle no. 1: the even numbered points of the original signal go in the first half of the array, and odd numbered points in the second half.

Now if we had G^* in the first half of the table and H^* in the second half, we could merge G_0^* and H_0^*, G_1^* and H_1^*, and so on, to obtain F^* in the whole table. Call this rearrangement "shuffle no. 1."

Now to calculate G^* by splitting G into two parts, we should move the even-numbered points of G into the first half of its table and the odd-numbered points into the second half of its table. The result of this "shuffle no. 2" is shown in Fig. 6.3.2.

Finally, "shuffle no. 3" produces 8 subtables of 1-point signals, as shown in Fig. 6.3.3.

Shuffle no. 1 put the even-numbered points in the first half of the original table and the odd-numbered points in the second half. What does this mean in terms of the index number of a typical point of F? Some thought shows that if the index number of a typical point of F is written in binary form, those with 0's in the *least* significant bit (even-numbered points) will be put in locations with 0's in the *most* significant bit (first half of table). Similarly, the second shuffle determines the *second-most* significant bit from the *second-least* significant bit.

The end result of all our rearrangement is to take F_k, *reverse the order of the bits in k,* producing the index k', and place F_k in position k'. This may be verified for $N = 8$ in Fig. 6.3.3, where the binary representations of the indices are shown. We shall call this process "bit reversal" or "shuffling."

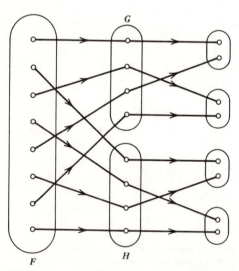

FIG. 6.3.2   Shuffle no. 2: the even-numbered points of G go in the first half of the G array, and the odd-numbered points in the second half of its array; and similarly for H.

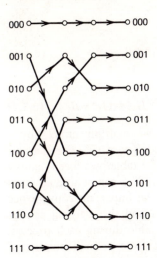

FIG. 6.3.3   The result of three stages of shuffling: bit reversal.

If this rearrangement process is performed *before we begin,* we can proceed by merging adjacent subtables in a straightforward way, finally producing F^* in the table at the last stage. Let us examine a typical merging step, say of two 4-element tables. This is shown in Fig. 6.3.4.

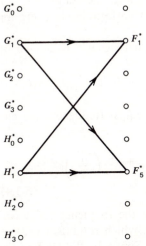

FIG. 6.3.4   A typical butterfly in the merging steps of the FFT algorithm.

We have shown the calculation

$$F_1{}^* = G_1{}^* + e^{-j2\pi/8}H_1{}^* = G_1{}^* + B \qquad (6.3.1)$$

and

$$F_5{}^* = G_5{}^* + e^{-j5\cdot2\pi/8}H_5{}^* = G_1{}^* - e^{-j2\pi/8}H_1{}^* = G_1{}^* - B \qquad (6.3.2)$$

In calculating the first half of the F^*–table, we multiply an H^* by a complex exponential, producing B, say, and add B to a G^*. For the corresponding spot in the second half of the F^*–table, we need simply $G^* - B$, and this saves half the multiplications. The computational steps indicated in Fig. 6.3.4 are called a "butterfly." An important convenience in the whole algorithm is that the bit reversal and the merging can be done "in place," without rewriting the whole table during each process. Thus the FFT requires N storage spaces, aside from a small number of auxiliary spaces.

Another practical consideration comes about when we wish to calculate the FFT repeatedly. It then may become economical to save the complex exponentials in a permanent table. This tradeoff of space for time may be of questionable value in some situations.

Figure 6.3.5 shows a complete FORTRAN program for calculating the FFT, written by Dr. Godfrey Winham. It is organized as a subroutine FTRANS that calls a subroutine SHUFF, which performs bit reversal; and a subroutine COMBIN, which does a merge operation. The computation leaves the original signal untouched and places the transform in a separate array, although we could have done the transform in place, erasing the original signal.

Figure 6.3.6 shows the output when the original signal S is defined by

$$S(n) = \sin\left(\frac{n\pi}{8}\right) \qquad n = 0, \ldots, N-1 \qquad (6.3.3)$$

a sine wave at a frequency of one-eighth the Nyquist frequency. We can predict the output by noting that

$$S(n) = \left(\frac{1}{2j}\right)(e^{jn\pi/8} - e^{-jn\pi/8}) = \left(-\frac{i}{2}\right)e^{jn\pi/8} + \left(\frac{i}{2}\right)e^{-jn\pi/8}$$

$$(6.3.4)$$

The DFT of the phasor $e^{jn\pi/8}$ is equal to N at the frequency point corresponding to one-eighth the Nyquist frequency, which is $(1/8)(N/2) = N/16$; and zero elsewhere (see Exercise 6.1.1). Similarly, the DFT of the phasor $e^{-jn\pi/8}$ is equal to N at the frequency point corresponding to

```
      DIMENSION S(1024)
      COMPLEX F(1024)
C        GENERATES TEST SIGNAL S AND OBTAINS DFT F USING FFT ALGORITHM
      N=32
      DO 1 J=1,N
    1 S(J)=SIN(FLOAT(J-1)*3.141593/8.)
      CALL FTRANS(S,F,N)
      DO 2 J=1,N
      FABS=CABS(F(J))
      JM=J-1
    2 WRITE(6,3)JM,S(J),F(J),FABS
    3 FORMAT(' ',I6,' SIGNAL=',F14.7,' F=',2F14.7,' FABS=',F14.7)
      STOP
      END

      SUBROUTINE FTRANS(S,F,N)
      DIMENSION S(1024)
      COMPLEX F(1024)
C        PLACES FOURIER TRANSFORM OF N-POINT SIGNAL S IN F
      CALL SHUFF(S,F,N)
      LENGTH=2
    1 DO 2 J=1,N,LENGTH
    2 CALL COMBIN(F,J,LENGTH)
      LENGTH=LENGTH+LENGTH
      IF (LENGTH-N) 1,1,3
    3 RETURN
      END

      SUBROUTINE SHUFF(S,F,N)
      DIMENSION S(1024)
      COMPLEX F(1024),CMPLX
C        'BIT-REVERSES' THE S ARRAY.  N (NO. OF POINTS) ANY POWER OF 2.
C            RESULT IS PUT IN F TO PREPARE TRANSFORM ITERATION
      DO 5 IFORT=1,N
      I=IFORT-1
      J=0
      M2=1
    1 M1=M2
      M2=M2+M2
      IF (MOD(I,M2)-M1) 3,2,2
    2 J=J+N/M2
    3 IF (M2-N) 1,4,4
    4 JFORT=J+1
    5 F(IFORT)=CMPLX(S(JFORT),0.)
      RETURN
      END

      SUBROUTINE COMBIN(F,J,N)
      COMPLEX F(1024),EMJT,Z,CEXP
C        COMBINES TRANSFORMS IN F(J)-F(N/2+J-1) AND F(N/2+J)-F(N+J-1)
C            INTO TRANSFORM IN F(J)-F(N+J-1)
      EMJT=CEXP((0.,-1.)*(6.283185/FLOAT(N)))
      N2=N/2
      DO 1 L=1,N2
      LOC1=L+J-1
      LOC2=LOC1+N2
      Z=EMJT**(L-1)*F(LOC2)
      F(LOC2)=F(LOC1)-Z
    1 F(LOC1)=F(LOC1)+Z
      RETURN
      END
```

FIG. 6.3.5   A FORTRAN program written by Godfrey Winham for the FFT algorithm. The main program sets up a test signal and calls SUBROUTINE FTRANS, which in turn calls SUBROUTINE's SHUFF and COMBIN.

0 SIGNAL=	0.0000000	P=	0.0000021	0.0000000 PABS=	0.0000021
1 SIGNAL=	0.3826835	P=	0.0000012	-0.0000029 PABS=	0.0000031
2 SIGNAL=	0.7071068	P=	0.0000023	-15.9999800 PABS=	15.9999800
3 SIGNAL=	0.9238793	P=	0.0000004	0.0000024 PABS=	0.0000025
4 SIGNAL=	0.9999999	P=	0.0000025	-0.0000000 PABS=	0.0000025
5 SIGNAL=	0.9238796	P=	-0.0000006	-0.0000003 PABS=	0.0000007
6 SIGNAL=	0.7071074	P=	0.0000031	0.0000070 PABS=	0.0000076
7 SIGNAL=	0.3826833	P=	-0.0000017	-0.0000003 PABS=	0.0000017
8 SIGNAL=	0.0000006	P=	0.0000003	-0.0000021 PABS=	0.0000022
9 SIGNAL=	-0.3826821	P=	-0.0000018	0.0000003 PABS=	0.0000019
10 SIGNAL=	-0.7071065	P=	-0.0000012	0.0000028 PABS=	0.0000030
11 SIGNAL=	-0.9238791	P=	-0.0000000	0.0000007 PABS=	0.0000017
12 SIGNAL=	-0.9999999	P=	-0.0000000	-0.0000057 PABS=	0.0000057
13 SIGNAL=	-0.9238798	P=	-0.0000008	0.0000009 PABS=	0.0000012
14 SIGNAL=	-0.7071065	P=	0.0000083	0.0000057 PABS=	0.0000101
15 SIGNAL=	-0.3826838	P=	-0.0000003	0.0000003 PABS=	0.0000005
16 SIGNAL=	0.0000007	P=	0.0000023	0.0000000 PABS=	0.0000023
17 SIGNAL=	0.3826833	P=	-0.0000003	-0.0000003 PABS=	0.0000005
18 SIGNAL=	0.7071061	P=	-0.0000015	-0.0000019 PABS=	0.0000024
19 SIGNAL=	0.9238796	P=	-0.0000008	-0.0000009 PABS=	0.0000012
20 SIGNAL=	0.9999999	P=	-0.0000000	0.0000057 PABS=	0.0000057
21 SIGNAL=	0.9238793	P=	-0.0000016	-0.0000007 PABS=	0.0000017
22 SIGNAL=	0.7071069	P=	-0.0000018	-0.0000012 PABS=	0.0000022
23 SIGNAL=	0.3826826	P=	-0.0000018	-0.0000003 PABS=	0.0000019
24 SIGNAL=	-0.0000000	P=	0.0000003	0.0000021 PABS=	0.0000022
25 SIGNAL=	-0.3826827	P=	-0.0000017	0.0000003 PABS=	0.0000017
26 SIGNAL=	-0.7071070	P=	0.0000019	-0.0000047 PABS=	0.0000051
27 SIGNAL=	-0.9238793	P=	-0.0000006	0.0000003 PABS=	0.0000007
28 SIGNAL=	-0.9999999	P=	0.0000025	0.0000000 PABS=	0.0000025
29 SIGNAL=	-0.9238796	P=	0.0000004	-0.0000024 PABS=	0.0000025
30 SIGNAL=	-0.7071061	P=	-0.0000108	15.9999700 PABS=	15.9999700
31 SIGNAL=	-0.3826832	P=	0.0000012	0.0000029 PABS=	0.0000031

FIG. 6.3.6 The output of the program in the previous figure for the sine-wave test signal of Eq. 6.3.3.

$- (1/8) (N/2) = -N/16$, which corresponds to $(15/16)N$. In our case, $N = 32$, so the DFT of (6.3.4) is

$$S^*(k) = \begin{cases} -j16 & \text{when} & k=2 \\ +j16 & \text{when} & k=30 \\ 0 & \text{elsewhere} \end{cases} \qquad (6.3.5)$$

which is verified in the output listing.

Exercise 6.3.1

Verify the fact that subroutine SHUFF bit reverses the S array. Can you think of another way to accomplish this?

Exercise 6.3.2 (computer experiment)

Change the given FORTRAN program so that it computes the FFT in place. Test your program by repeatedly transforming a given signal 4 times (see Exercise 6.1.3).

Exercise 6.3.3 (computer experiment)

Run an FFT program for a sequence of values of N, say $N = 8, 16,$ $\ldots, 1024$, and plot the execution time reported. Explain any discrepancies with theory.

Exercise 6.3.4

Step through by hand the FFT algorithm for $N = 4$ and an arbitrary input sequence $F_0, \ldots, F_3$. Verify that your results are the same as the DFT definition, Eq. 6.1.5.

6.4 A FURTHER PROPERTY OF THE DFT: CIRCULAR CONVOLUTION

We saw in Section 5.5 that the product of two z-transforms, say $X^*(z)$ and $Y^*(z)$, was the z-transform of a third signal, called the convolution of $X(k)$ and $Y(k)$. More specifically, we derived the following z-transform:

$$X(k) \otimes Y(k) = \sum_{j=0}^{\infty} X(j) Y(k-j) \xrightarrow{Z} X^*(z) Y^*(z)$$

$$(6.4.1)$$

We would expect, therefore, that if we took the DFT of two finite length signals, and multiplied the results together at corresponding frequency points, that the result would be close to the DFT of the convolution of the original two signals. To find out the exact results, we need to calculate the signal $Q(m)$ obtained by the following operations, where $X(k)$, $Y(k)$, and $Q(m)$ are all digital signals of length N:

$$Q(m) = \text{IDFT}\{\text{DFT}[X(k)] \cdot \text{DFT}[Y(k)]\} \qquad (6.4.2)$$

This signal is the counterpart of the convolution of the signals $X(k)$ and $Y(k)$, except that the DFT operation has replaced the z-transform operation.

Substituting

$$X^*(n) = \text{DFT}[X(k)] = \sum_{k=0}^{N-1} X(k) e^{-jkn2\pi/N} \qquad (6.4.3)$$

and

$$Y^*(n) = \text{DFT}[Y(h)] = \sum_{h=0}^{N-1} Y(h) e^{-jhn2\pi/N} \qquad (6.4.4)$$

into (6.4.2), we get

$$Q(m) = \frac{1}{N} \sum_{n=0}^{N-1} e^{+jnm2\pi/N} \sum_{k=0}^{N-1} X(k) e^{-jkn2\pi/N} \sum_{h=0}^{N-1} Y(h) e^{-jhn2\pi/N}$$

$$(6.4.5)$$

Rearranging the order of these finite summations yields

$$Q(m) = \frac{1}{N} \sum_{k=0}^{N-1} \sum_{h=0}^{N-1} X(k) Y(h) \left[\sum_{n=0}^{N-1} e^{jn(m-k-h)2\pi/N} \right] \qquad (6.4.6)$$

We have already encountered the summation in brackets; it is the same as (6.1.10), except that $(m - k - h)$ appears in place of $(p - n)$. The same sort of argument as was used before leads to the following result:

$$\sum_{n=0}^{N-1} e^{jn(m-k-h)2\pi/N} = \begin{cases} N & \text{if} \quad m-k-h=0 \,(\text{mod } N) \\ 0 & \text{otherwise} \end{cases}$$

$$(6.4.7)$$

The condition $m - k - h = 0$ (mod N) is the same as $h = m - k$ (mod N), and only these values of h contribute nonzero terms in (6.4.6). Hence, (6.4.6) finally becomes

$$Q(m) = \sum_{k=0}^{N-1} X(k) \cdot Y[(m-k) \,(\text{mod } N)] \qquad (6.4.8)$$

This is exactly the same as ordinary convolution (5.5.2) except for one detail: the N points are treated as being wrapped around a circle so that the $(N-1)$st sample immediately precedes the zeroth sample. Hence the operation is called *circular* convolution. Figure 6.4.1 illustrates this difference.

In practice, circular convolution can be used to calculate the ordinary convolution, provided the finite length signals are "padded" with zeros to eliminate the effect of the wrapped-around "tails," such as in Fig. 6.4.1*d*. For more details the reader is referred to the chapter by T. G. Stockham, Jr. in reference 1 at the end of this chapter.

Exercise 6.4.1

Consider the situation where we wish to filter a signal with N sample points using a moving average filter having M coefficients, where M is large. Instead of direct filtering, we could take the DFT of the signal,

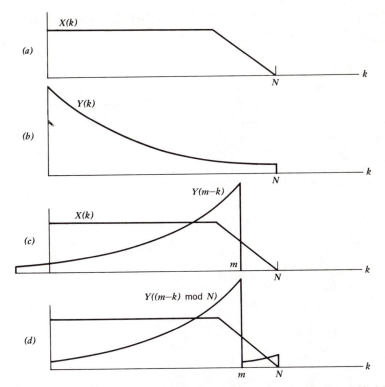

FIG. 6.4.1 Ordinary and circular convolution of the signals X and Y; (a) original signal $X(k)$; (b) original signal $Y(k)$; (c) ordinary convolution; (d) circular convolution.

multiply the value at each frequency point by the value of the DFT of the filter coefficients at the same point, and then take the IDFT of the result. Discuss the relative efficiency of this procedure for different M and N, assuming that the FFT algorithm is used to calculate the DFT and IDFT. Also assume that the "wrap-around" effect has been taken care of.

Exercise 6.4.2

Use the circular convolution formula (6.4.8) to show that

$$\mathrm{DFT}[F((k-p)\,(\mathrm{mod}\,N))] = F^*(n)\,e^{-jnp2\pi/N}$$

where k is considered to be the sample number on the left, and n on the right. This is analogous to property 1 of z-transforms.

6.5 EXAMPLE: APPLICATION TO SPEECH ANALYSIS AND SYNTHESIS

The ideas of digital signal processing that we have discussed in Part I have had a profound effect on our ability to analyze and synthesize human speech. The advent of fast and inexpensive digital hardware makes possible the construction of flexible and complex systems for sampling speech waveforms, obtaining efficient and illuminating representations of speech sounds, and synthesizing intelligible speech automatically. We outline below some of the basic ideas in this work.

Figure 6.5.1 shows a schematic diagram of the human vocal tract.

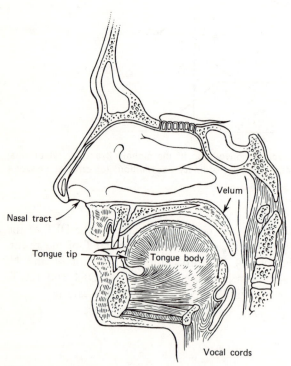

FIG. 6.5.1   Schematic diagram of the human vocal tract.

Voiced sounds (such as "AH," "EE," etc.) are produced by periodic vibration of the vocal cords, sending periodic pulses of air pressure through the vocal tract, which acts as a linear filter. Fricative sounds (such as "SHHH,"

"SSS," etc.) are produced by forcing air through a constriction somewhere in the vocal tract, thereby producing turbulent pressure waves that travel through the rest of the vocal tract. Plosive sounds (such as "K," "D," etc.) are produced by air pressure waves produced by a sudden release of pressure somewhere in the tract.

This system can be simulated by the digital filter system shown in Fig. 6.5.2.

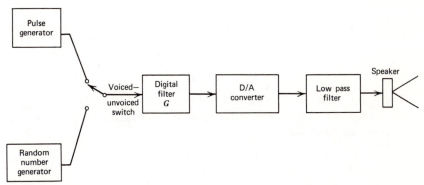

FIG. 6.5.2   Digital filter system that simulates the human vocal tract.

The digital filter G represents the effect of the shape of the vocal system at any time, and determines which particular vowel or consonant is produced. Thus, during operation, G must be changed often (every 10 msec is adequate) to reflect the changes in the vocal system during real speech.

The digital filters G corresponding to various vowels and consonants have been determined experimentally; they have peaks in their frequency responses corresponding to resonances in the vocal tract in various stages of speaking. Cascade chains of four or five two-pole recursive filters are commonly used in hardware implementations, called "digital vocoders."

The FFT is used widely for analyzing speech, and the results are often displayed in the form of a spectrogram, which is obtained as follows. The sampled speech signal is broken down into short segments (say 10 msec), over which intervals the frequency content is assumed to be nearly constant. The FFT is applied to each segment, and the results are plotted as a two-dimensional picture with time running from left to right; frequency running from bottom to top; and frequency content represented by variations in darkness, the darker areas representing larger frequency content. An example is shown in Fig. 6.5.3, with the words shown above the time axis.

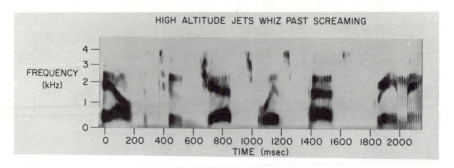

FIG. 6.5.3 Spectrogram of the utterance "HIGH ALTITUDE JETS WHIZ PAST SCREAMING." (Photograph courtesy of Dr. L. R. Rabiner, Bell Telephone Laboratories.)

The reader interested in pursuing further the applications of digital signal processing to speech is referred to the references on speech in the suggestions for further reading.

6.6 TRANSITION

The topics we have studied until now comprise the elements of what might be called "linear system theory." The central notions are those of phasor, frequency response, transforms, the complex frequency plane, recursive filters, poles and zeros, partial fraction expansion, and so on. Everything we have done has been in terms of discrete-time (digital) signals, and any filters that need to be realized would be realized by a digital computer algorithm or constructed from computer components.

At this point we could take up the corresponding theory of continuous-time (analog) signals and filters. Many of the ideas developed thus far have direct counterparts in analog systems and these ideas will hopefully facilitate the study of such systems. For example, without going into details, consider an electrical network composed of resistors, capacitors, and inductors; with a continuous-time sine-wave generator applied to certain (input) terminals, and a voltmeter connected to certain other (output) terminals (Fig. 6.6.1). Such a system has profound analogies with a digital filter and can be analyzed in much the same way. The electrical network has a transfer function (with poles and zeros), and a frequency response. The steady state output with a sinusoidal input will be a sinusoid, with amplitude and phase determined by the transfer function of the network.

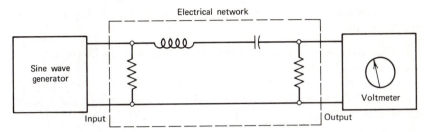

FIG. 6.6.1   A continuous-time filter.

The focal point of this book is the use of the computer, however, and we choose therefore to change direction, and to take up the study of algorithms with different kinds of systems in mind—systems described by *graphs*. Chapters 7 to 11 will deal therefore with simple and fundamental properties of systems described by points (nodes) connected by lines (branches). We shall consider some problems associated with such systems, and shall illustrate the power of computer algorithms to deal effectively with such problems. As we shall see, the idea of a graph is so fundamental that many diverse situations can be described in terms of graphs.

Finally, in Chapter 11, we shall return to electrical networks such as the one illustrated in Fig. 6.6.1, which is, after all, a graph with points (electrical terminals) connected by lines (electrical components). It is hoped that this tour will prepare the reader for further studies in the rather diverse fields of engineering and computer science.

Further Reading

The following book will provide further reading on many of the subjects discussed up to this point:

1. *Digital Processing of Signals,* B. Gold and C. M. Rader, McGraw-Hill, New York, 1969.

The z-transform is treated in detail in references 2 to 4.

2. *Sampled-Data Control Systems,* J. R. Ragazzini and G. F. Franklin, McGraw-Hill, New York, 1958.

3. *Theory and Application of the z-Transform Method,* E. I. Jury, Wiley, New York, 1964.

4. *Discrete-Time Systems,* H. Freeman, Wiley, New York, 1965.

Those interested in pursuing recent work in digital filtering and the Fast Fourier Transform will find ample material in the following special issues in reference 5. References 6 and 7 may also prove helpful.

5. *IEEE Transactions on Audio and Electroacoustics,* vol. AU-15, no. 2, June 1967, Special Issue on the Fast Fourier Transform and its Application to Digital Filtering and Spectral Analysis; vol. AU-17, no. 2, June 1969, Special Issue on the Fast Fourier Transform; vol. AU-18, no. 2, June 1970, Special Issue on Digital Filtering; vol. AU-18, no. 4, December 1970, Special Issue on Digital Signal Processing; vol. AU-20, no. 4, October 1972, Special Issue on Digital Filtering.

6. *Papers on Digital Signal Processing,* A. V. Oppenheim (editor), MIT Press, Cambridge, Mass., 1969.

7. *Digital Signal Processing,* L. R. Rabiner and C. M. Rader (editors), IEEE Press, New York, 1972.

The ancient history (back to C. Runge in 1903) of the FFT is described in reference 8.

8. "Historical Notes on the Fast Fourier Transform," J. W. Cooley, P. A. W. Lewis, and P. D. Welch, *IEEE Trans. on Audio and Electroacoustics,* vol. AU-15, no. 2, June 1967, pp. 76–79.

A general reference on speech is the following.

9. *Speech Analysis, Synthesis and Perception,* J. L. Flanagan, Springer-Verlag, New York, 1971 (2nd edition).

A survey of the applications of digital signal processing, with quite a few references, can be found in reference 10.

10. "A Survey of Digital Speech Processing Techniques," R. W. Schafer, *IEEE Trans. on Audio and Electroacoustics,* vol. AU-20, no. 1, March 1972, pp. 28–35.

The following papers (references 11 and 12) report work on the production of synthetic speech.

11. "Synthetic Voices for Computers," J. L. Flanagan, C. H. Coker,

L. R. Rabiner, R. W. Schafer, and N. Umeda, *IEEE Spectrum,* vol. 7, October 1970, pp. 22–45.

12. "The Synthesis of Speech," J. L. Flanagan, *Scientific American,* vol. 226, no. 2, February 1972, pp. 48–58.

A test for the stability of a linear time-invariant digital filter, which does not require factoring a polynomial, is described in reference 13.

13. "A Stability Test for Linear Discrete Systems Using a Simple Division," E. I. Jury, *Proc. IRE* (Letters), vol. 49, no. 12, December 1961, p. 1947.

Other methods for calculating the inverse z-transform are described in references 14 and 15.

14. "A New Method of Obtaining Inverse z-Transforms," L. R. Badgett, *Proc. IEEE* (Letters), vol. 54, no. 7, July 1966, p. 1010.

15. "A Determinant Formulation for the Inverse z-Transform," N. Ahmed and K. R. Rao, *Proc. IEEE* (Letters), vol. 55, no. 11, November 1967, p. 2031.

GRAPHS
AND ALGORITHMS

7.
GRAPHS AND
THEIR STORAGE

7.1 INTRODUCTION

In Part I we studied techniques for processing signals, and the relevant mathematical theory was that of complex numbers and functions. We turn now to a second class of engineering problems, those that deal with systems of interconnected components, and the relevant mathematics is called *graph theory*. The concepts of graph theory that we require will be quite simple; simpler, for example, than the notions of frequency response or convolution. The ideas will have a characteristic "yes-no" flavor, and for the most part no complicated arithmetic operations will be involved. This simplicity has important implications with regard to the usefulness of the subject: first, the manipulations involved are ideally suited to the use of digital computers, and we shall find that very large and complex systems can be investigated efficiently by straightforward programs. Second, the notions used will correspond very directly with the real world. It is often quite simple, for example, to build a believable model for a system in terms of graphs. This accounts in large part for the growing importance of graph theory in dealing with problems of organizing interacting parts of a complex world. Presently, the techniques are used to study networks of telephones, satellites, computers, roads, pipelines, production facilities, and many other components of technology.

One way of looking at our earlier study of signals is as a microscopic view of information processing. We have control, when dealing with a particular digital filter, over each bit that passes our way. Suppose, however, that we wish to build a network of communications facilities that interconnects many cities. We then are forced to step back and take a broader view of things. We may regard one city as a point connected to other points, and study the flow of information in and out of each city.

This is the viewpoint of system theory: macroscopic rather than microscopic.

It is appropriate at this point to define what we mean by the term graph. By a *graph* we mean a collection of points, called *nodes,* some of which may be interconnected by lines, called *branches.* This is a different meaning of the term from that used in analytical geometry, where a graph is a plot of function values. There is no harm done by this ambiguity, since there is no danger of confusing these two meanings. While on the subject of terminology, we should mention that nodes are sometimes called *vertices,* and branches are sometimes called *edges* or *arcs.*

More abstractly, we can define a graph as two sets: a set N of nodes, and a set B of branches. The set B contains certain unordered pairs of nodes: those pairs connected by branches. Thus, the mathematical notation for a graph G is

$$G = \{N, B\} \tag{7.1.1}$$

a pair of sets. As an illustration, consider the graph shown in Fig. 7.1.1.

FIG. 7.1.1   Example of a graph.

This graph has 4 nodes, labeled 1 through 4, and 3 branches. The mathematical notation for this graph is then

$$G = \{N, B\}$$

where

$$N = \{1,2,3,4\}$$

and

$$B = \{(1,2), (2,3), (1,4)\} \tag{7.1.2}$$

Although we shall usually represent a graph by a picture, or by some arrays of numbers in a computer, this mathematical notation illustrates some important points. First, the sets of nodes and branches are not in any particular order. Thus, we could also have written

$$N = \{3,2,1,4\}$$

and

$$B = \{ (2,3), (1,4), (1,2) \} \tag{7.1.3}$$

or any other such rearrangement. Second, the pair of nodes representing any particular branch is unordered, so that we could write as well

$$B = \{ (3,2), (4,1), (1,2) \} \tag{7.1.4}$$

This means that there is no particular direction associated with a branch, and the term *undirected graph* is sometimes used to emphasize this. Later on, we shall want to consider graphs whose branches do have directions associated with them, and these will be called *directed graphs*.

There are two special situations that might arise. First, a node may be connected to itself. This corresponds to a branch of the form (a,a), which connects node a with itself. Such a branch is called a *self-loop*. Second, there may be more than one branch connecting a particular pair of nodes. This corresponds to a set B like the following:

$$B = \{ \ldots, (a,b), (a,b), \ldots \} \tag{7.1.5}$$

Such branches are called *multiple branches*. Generally, we shall find it convenient to assume that graphs have neither self-loops nor multiple branches.

We have not said anything about where the nodes are to be placed when we draw a graph, or what shape lines are to be used to indicate branches. These decisions are determined by convenience and do not affect the identity of the graph being represented. For example, we could redraw the graph of Fig. 7.1.1 without any branches crossing, as shown in Fig. 7.1.2. If it is possible to draw a graph on a plane sheet of paper so that no two branches cross, the graph is called *planar*. It often clarifies the picture of a graph if we can draw it so as to minimize the number of branch crossings. On the other hand, the nodes might represent cities on a map, in which case we might want to preserve their relative positions to indicate distances between them. Thus we think of a graph abstractly as being a set of nodes and a set of branches, and we draw a picture of a graph so it is easy to read.

FIG. 7.1.2 A redrawing of the graph shown in the previous figure.

Exercise 7.1.1

Each of three houses needs to be connected to sources of gas, water, and electricity, each of which is in a separate building. Describe this situation by a graph, using abstract notation. Draw a picture of the graph. Convince yourself that it is not possible to draw this graph on a plane sheet of paper without crossing branches. (Such graphs are called *nonplanar.*)

Exercise 7.1.2

What is the maximum number of branches that a graph may have, excluding self-loops and multiple branches? (Graphs with this property are called *complete.*)

Exercise 7.1.3

What is the minimum number of branches that a graph may have, assuming that every node is connected to at least one other node?

7.2 SOME EXAMPLES OF GRAPHS

We now take up some examples of graphs that occur in engineering problems. As a first example we shall consider a communication network; that is, a graph whose branches represent communication links between the nodes. Figure 7.2.1 shows such a graph with 58 nodes. The nodes, or "cities" as we may think of them, are placed in a regular grid that might represent a geographical distribution of communication centers. Notice that each node is connected to exactly 6 other nodes, so that the total number of branches is $(58)(6)/2 = 174$. This is true because if we multiply the number of nodes times the number of branches connected to each node, we count each branch exactly twice. This graph has an interesting property, one that is not at all obvious from its picture: the removal of any 5 nodes is not sufficient to break the graph into disconnected pieces. In other words, if any 5 nodes are removed, it is still possible to find a path of communication between any two remaining nodes. This property means that the communication network has a certain amount of "redundancy" and will remain useful even after some of the nodes or

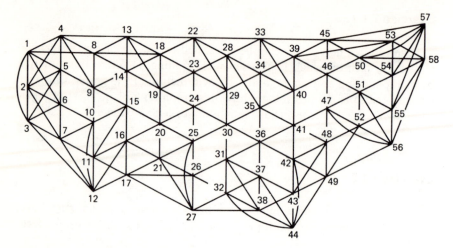

FIG. 7.2.1 A graph that represents a communication network; the removal of no 5 nodes is sufficient to break the graph into disconnected pieces.

branches fail or are destroyed. The reader may find it amusing to try to find a set of less than 6 nodes whose removal breaks the graph into 2 or more disconnected pieces.

Communication networks, as examples of graphs, are useful to keep in mind as we proceed. They have the advantage of being concrete, since we can think of the nodes as points on a geographical map, and the branches as telephone lines or radio links.

As another example of a graph, one that is not physical in nature, we refer to the FFT algorithm discussed in Chapter 6. Figure 7.2.2 shows a graph that illustrates the successive merging steps described in Sections 6.2 and 6.3. The rightmost column represents one 8-point transform; the second column from the right represents two 4-point transforms; the next column represents four 2-point transforms; and the leftmost column represents eight 1-point transforms. As we proceed from left to right, a branch is drawn whenever a point at the left is used in the calculation of a point at the right. Since each point is used in the calculation of two points in the next stage, and since the calculation of each point uses two points from the previous stage, each node has four branches connected to it (except for the first and last stages).

As a third example of a graph, we consider the collection of natural gas from offshore drilling platforms, and the transmission of the gas through pipelines to an onshore collection point. Figure 7.2.3 shows a graph that represents such a system. Each node except the topmost one represents a drilling platform and each branch represents an underwater pipeline. The

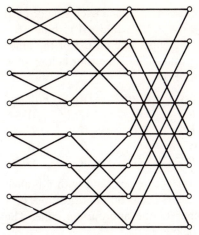

FIG. 7.2.2 A graph that represents the computation steps in the FFT algorithm.

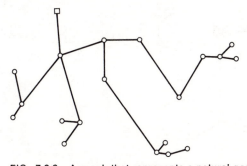

FIG. 7.2.3 A graph that represents a natural gas pipeline system.

topmost node represents the onshore collection point for the natural gas. Typical offshore gas pipeline systems can be found in the Gulf of Mexico, off the coast of Louisiana. Such systems cost many millions of dollars, and sophisticated methods have been developed for designing them efficiently.

Example*

Sometimes, merely drawing the right graph helps a great deal towards solving a problem. Consider the puzzle marketed by the name "Instant

* See references 2 and 4 in the suggestions for further reading at the end of this chapter.

Insanity": we are given four cubes, say cubes 1 to 4, and each side of each cube is colored one of four colors, say red (R), white (W), blue (B), and green (G). The problem is to stack the cubes one on top of the other so that each side of four faces has exactly one each of the four colors. One assignment of colors to cube faces is given below.

Cube number	North	South	East	West	Top	Bottom
1	R	R	R	W	W	W
2	B	W	B	G	G	G
3	R	W	W	G	R	B
4	B	W	B	G	B	G

Draw the graph G with four nodes, one for each color; and with a branch between two colors whenever those two colors occur on opposite sides of a cube. Label each branch with the number of the appropriate cube. The graph G for the problem above is drawn below.

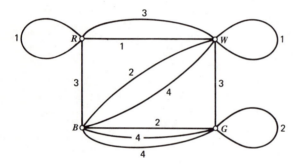

Now suppose we have a solution, and consider first the east-west faces. The four pairs of faces represent four branches, one from each cube. Furthermore, each color is represented exactly twice. Thus, the east-west face pairs are represented by a graph with branches taken from G, with the property that each cube is represented by one branch, and each node has degree two. (See Exercise 7.2.2 for the definition of degree.) We call a graph with branches taken from G a *subgraph* of G. The east-west face pairs then determine a subgraph of G with four branches, each from a different cube, so that each node is of degree two.

Similarly, the north-south face pairs determine another such subgraph, with branches disjoint from the east-west subgraph.

To summarize, a solution to the problem determines two disjoint subgraphs with the properties:

(1) Each subgraph has exactly one branch from each cube.

(2) Each node is of degree two.

Furthermore, each such pair of disjoint subgraphs determines a solution, since the cubes can first be stacked so that the east-west face pairs are determined by the first subgraph, and then each cube can be rotated about its east-west axis so that the north-south face pairs are determined by the second subgraph.

Two subgraphs that determine a solution for the problem above are shown below:

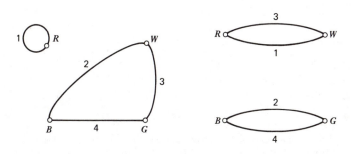

Example (coloring graphs)*

The problem of collecting garbage in a city provides us with another application of graph theory. Suppose we have n locations in a city, each requiring garbage collection. Represent these locations by n nodes as shown below:

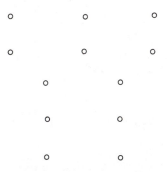

On a given day we can send out a crew of men with a truck to collect garbage from a number of locations, say up to four locations. Each day's

---

* This example was suggested by Professor Alan C. Tucker.

work for a crew can be represented by a tour of up to four nodes. Five
typical tours are shown below, labeled T_1 to T_5.

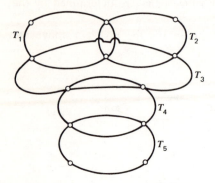

One problem that arises is the following: certain locations are visited by
more than one tour, and the tours should be scheduled on days of the
week so that no location is visited more than once on the same day. To
deal with this problem it is convenient to draw a new graph, called a "tour
graph" and denoted by G_T, as follows:

(1) Nodes of G_T correspond to the tours $T_1, \ldots, T_5$;

(2) Two nodes of G_T are connected by a branch whenever the two
     corresponding tours both visit the same location.

The tour graph G_T for the example is shown below:

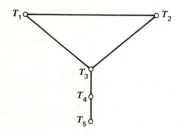

The problem of scheduling tours on days of the week can now be stated
in graph theory terms as:
       Assign days of the week to nodes of the tour graph so that no two
       connected nodes are assigned the same day.
It is easy to see that we must send crews out on at least three different days

to satisfy the requirements of our example. Denoting Monday, Tuesday, and Wednesday by *M, T,* and *W,* one solution is shown below:

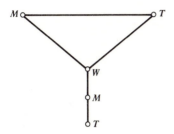

The same sort of problem comes up in coloring maps: a cartographer must assign colors to countries (or states) so that no two bordering countries are assigned the same color. Draw a graph in which each node represents a country, and connect two nodes whenever the two corresponding countries share a border. The graph theory problem then becomes one of assigning colors to nodes so that no two connected nodes have the same color. A graph corresponding to some states of the United States is shown below:

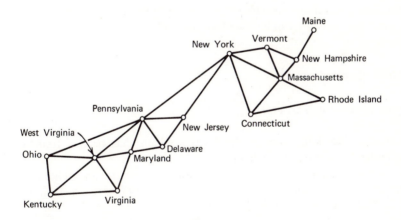

A coloring with four colors, 1,2,3, and 4, is shown on the next page.

Because of this application, the problem of labeling nodes of a graph so that no connected nodes have the same label is called the "graph coloring problem." The following terms are used:

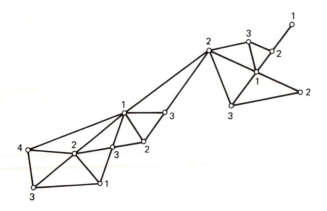

Definition

A graph G is *p-colorable* if its nodes can be colored with p colors so that no connected nodes have the same color. The smallest p such that G is *p*-colorable is called the *chromatic number* $\chi(G)$ of G.

It has been proved that for a planar graph G (and hence for the map-coloring problem, see Exercise 7.2.11) $\chi(G) \leq 5$. Although it appears that only 4 colors are sufficient to color any planar graph, the best attempts of great mathematicians for almost a century have failed to produce a proof. The open question can be stated as follows:

Four Color Conjecture

For any planar graph $G, \chi(G) \leq 4$.

The problem, like other famous unsolved problems, has been a tremendous stimulant to research in its field.

Two other problems in which graph coloring arises are mentioned by Christofides ("An Algorithm for the Chromatic Number of a Graph," N. Christofides, *Computer Journal, 1971*):

Problem

Chemicals must be stored in rooms and certain pairs of chemicals, for safety reasons, cannot be stored in the same room. Store the chemicals in the smallest number of rooms.

Problem

Examinations must be scheduled and certain students must take more than one examination. Schedule the examinations in the smallest number of periods so that no two examinations to be taken by the same student occur simultaneously.

Exercise 7.2.1

Consider the graph like Fig. 7.2.2 associated with an N-point transform. How many branches are there in such a graph? Exactly half of these branches correspond to multiplications in the FFT algorithm.

Exercise 7.2.2

The *degree* $D(i)$ of the node i of a graph is the number of branches attached to node i, counting self-loops as two. Show that the total number of branches in a graph is given by the formula

$$\left(\frac{1}{2} \right) \sum_{i=1}^{N} D(i)$$

where N is the number of nodes.

Exercise 7.2.3

Associate one node with each of the 64 squares of the chessboard, and put a branch between two nodes if it is possible for a Queen to move from the square associated with one node to the square associated with the other in one move. Which nodes have the highest degree? What is that degree? Which nodes have the lowest degree? What is that degree? Repeat for the Rook, Bishop, and Knight.

Exercise 7.2.4

Show that in any graph the number of nodes whose degree is odd, is even.

Exercise 7.2.5

Consider the problem of drawing a graph without lifting one's pencil from the sheet of paper, without retracing a branch, but possibly passing through nodes more than once. Discuss the requirements on the degrees of the nodes for this to be possible. (Graphs that can be drawn in this way were first studied by Euler, whose name we encountered earlier in connection with complex numbers.)

Exercise 7.2.6

Make up three situations, other than those already mentioned, that can be described by graphs, and provide a concrete example of each.

Exercise 7.2.7

Consider another problem posed by four cubes colored with four colors: stack the cubes so that each side of four faces is colored all red, all white, all blue, and all green. (Call this problem "Instant Sanity.") Show how to solve this problem using the graph G. Prove that if a solution to Instant Sanity exists, then a solution to Instant Insanity exists, and show how to obtain the latter from the former. Is the converse true? That is, if a solution to Instant Insanity exists, does a solution to Instant Sanity exist?

Exercise 7.2.8

Prove that the chromatic number of the graph given above representing part of the United States is 4.

Exercise 7.2.9

Prove that the chromatic number of a tree is 2. (See Section 7.4 for the definition of a tree.)

Exercise 7.2.10

What is the chromatic number of a complete graph with n nodes (see Exercise 7.1.2) ?

Exercise 7.2.11

Let M be a graph whose branches represent the borders of a map drawn in the plane. Define the *dual* of M, M^*, to be the graph constructed by placing a node in each country, and connecting nodes of M^* whenever the corresponding countries share a border. Examples of M and M^* are shown below:

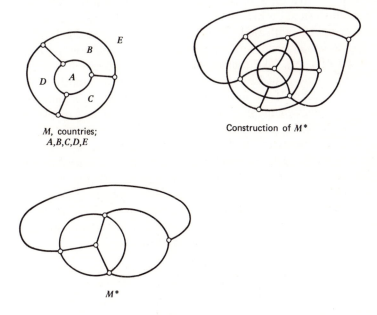

M, countries;
A,B,C,D,E

Construction of *M**

M*

Prove that if M corresponds to a map drawn in the plane, M^* is planar.

7.3 THE STORAGE OF GRAPHS IN A COMPUTER

Since we shall often want to deal with graphs in digital computer programs, we now shall discuss how the information describing a graph can be stored. There are many ways to do this, and the subject is by no means trivial, since certain methods of storage are particularly efficient for certain kinds of calculations. Three methods will be described below, but first we need to introduce another term. If a node j is connected to a given node i, we say that j is *adjacent* to node i. Thus, node 1 in Fig. 7.1.1 is adjacent to node 2 but not to node 3.

A. The Branch-List Method

This method corresponds directly to the mathematical notation introduced in Section 7.1. We think of the branches as being in a list. We then define two one-dimensional arrays, B1(I) and B2(I) as follows: if the Ith branch connects node j with node k, we set B1(I) $= j$ and B2(I) $= k$. Notice that the order of j and k is immaterial, and we could as well set B1(I) $= k$ and B2(I) $= j$. The dimension of the one-dimensional arrays B1(I) and B2(I) is determined by the maximum number of branches that are allowed in the graph. If we have no information about this limit, we should allow for a maximum of $n(n-1)/2$ branches to be completely safe (see Exercise 7.1.2).

To illustrate the branch-list method of storage, consider the following example of a graph: a tournament takes place in which certain participants play games with certain other participants. Represent each player by a node, and represent the fact that player i played player j by the branch (i,j). Figure 7.3.1 shows such a graph in the case where 6 players each played 3 others. There are 9 branches in this graph, and the corresponding branch-list representation is shown below:

I	B1(I)	B2(I)
1	1	2
2	1	3
3	1	4
4	2	6
5	2	5
6	3	5
7	3	6
8	4	5
9	4	6

(7.3.1)

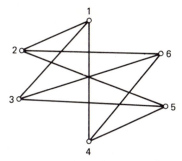

FIG. 7.3.1   A graph representing a tournament.

B. The Adjacency-List Method

In this method, two arrays are used, the first one-dimensional and second two-dimensional. The first array, which we shall call NADJ(I) has in its Ith position the number of nodes adjacent to node I. The second array, called NEAR(I,J), has in its (I,J)th position the Jth node that is adjacent to node I. The NADJ(I) adjacent nodes are stored in any order, but without repetition. Thus, NADJ(I) tells us how far down the NEAR(I,J) list to look to find all the nodes adjacent to node I. To illustrate this the tournament graph of Fig. 7.3.1 would be stored as follows:

I	NADJ(I)	NEAR(I,J), J = 1	2	3
1	3	2	3	4
2	3	1	6	5
3	3	1	6	5
4	3	1	6	5
5	3	2	3	4
6	3	2	3	4

$$(7.3.2)$$

Notice that with the adjacency-list method it is particularly convenient to enumerate the nodes adjacent to a given node. If we wish to print every branch in a graph with N nodes, for example, we could use the following fragment of FORTRAN code:

```
INTEGER NADJ(50),NEAR(50,50)
.
.
.
DO 1 I=1,N
L=NADJ(I)
IF(L.EQ.0)GOTO 1
DO 2 J=1,L
2 WRITE(6,3)I,NEAR(I,J)
3 FORMAT(' BRANCH FROM',I4, ' TO', I4)
1 CONTINUE
.
.
.
```

Note that this code will print each branch twice; once for each node to which it is connected. One way to print each branch once and only once is to put in a test so that branch (I,J) is printed only when J > I.

On the other hand, it is a bit more complicated to find out if a node J is adjacent to node I, since we have to search the list of nodes adjacent to node I for J. If FLAG is a logical variable indicating whether J is adjacent to I, the following fragment of code will suffice: (i.e., the result will be FLAG = .TRUE. if J is adjacent to I, and FLAG = .FALSE. otherwise)

```
INTEGER NADJ(50),NEAR(50,50)
LOGICAL FLAG
.
.
.
FLAG=.FALSE.
L=NADJ(I)
IF(L.EQ.0)GOTO 1
DO 2 K=1,L
IF(NEAR(I,K).EQ.J)FLAG=.TRUE.
2 CONTINUE
1 CONTINUE
.
.
.
```

This problem of finding out whether a branch is present in a graph is even more cumbersome with the branch-list representation, since the entire list of branches must be searched.

C. The Adjacency-Matrix Method

The last method requires one two-dimensional array, which we shall call ADJ(I,J). ADJ(I,J) has in its (I,J)th place a "1" if nodes I and J are adjacent, and a "0" otherwise. Thus, the graph discussed above would be stored as follows:

ADJ(I,J)	I	J= 1	2	3	4	5	6
	1	0	1	1	1	0	0
	2	1	0	0	0	1	1
	3	1	0	0	0	1	1
	4	1	0	0	0	1	1
	5	0	1	1	1	0	0
	6	0	1	1	1	0	0

$$(7.3.3)$$

To print every branch in a graph, we now would use code that searched every element of the ADJ(I,J) matrix:

```
INTEGER ADJ(50,50)
   .
   .
   .
   DO 1 I=1,N
   DO 1 J=1,N
   IF(ADJ(I,J).EQ.1)WRITE(6,2)I,J
 1 CONTINUE
 2 FORMAT(' BRANCH FROM',14,' TO',14)
   .
   .
   .
```

Again, we could suppress duplication by printing a branch only when J > I.
With the adjacency-matrix method it is very simple to determine whether
a branch is in a given graph, since we can test ADJ(I,J) directly, as the
following code fragment illustrates:

```
INTEGER  ADJ(50,50)
.
.
.
FLAG = .FALSE.
IF(ADJ(I,J).EQ.1)FLAG = .TRUE.
.
.
.
```

The choice of form of graph storage for a particular application depends on several factors. One factor, as we have seen above, is the ease
with which certain searches can be performed. For example, if we know
that we need to search many times from given nodes to adjacent nodes, it
is logical to use the adjacency-list method. Another consideration is the
total amount of storage that the particular methods require. Suppose, for
example, that we wish to store a graph with 1000 nodes, but we know
that there can be at most 2000 branches. Then the branch-list method will
require 4000 words of reserved storage. On the other hand, the adjacency-
matrix method will require the storage of a matrix with dimensions 1000 ×
1000 = a million words, which might be impossible in a given program.

Exercise 7.3.1

Invent a fourth method for storing graphs in a computer.

Exercise 7.3.2

Number the nodes of the graph shown in Fig. 7.2.3 and write the
storage arrays for this graph for all 3 methods.

Exercise 7.3.3

We have described three methods, A, B, and C, for representing graphs in a computer. Suppose we wish to write a program that will convert any of these forms to any of the others. Discuss the representation of this problem by a graph. In particular, decide whether you want to use branches which have directions associated with them.

Exercise 7.3.4

Determine the maximum storage requirements of each of the graph representation methods.

Exercise 7.3.5

How can the graph storage methods discussed be modified to take into account the possibility of multiple branches between pairs of nodes?

Exercise 7.3.6

Repeat Exercise 7.3.5 for the possibility of self-loops.

Exercise 7.3.7

An n-cube is a graph defined as follows: There are 2^n nodes, v_0, v_1, . . . , v_{2^n-1}. The node v_i is associated with the binary representation of the number i. Thus, the 3-cube has 8 nodes, associated with the binary numbers $000, 001$, . . . ,111. Two nodes v_i and v_j are connected by a branch if and only if the associated binary numbers i and j differ in only one place. Draw pictures of the 1-,2-,3-, and 4-cubes. What is the degree of a node in an n-cube? How many branches are there in an n-cube?

Exercise 7.3.8

A common paperweight consists of a dodecahedron (a solid with 12 plane faces) with a month on each face. If each vertex of the dodecahedron is associated with a node of a graph, and each edge with a branch, draw a picture of such a calendar graph on a plane sheet of paper. Is it possible to arrange the months so that consecutive months always share an edge?

Exercise 7.3.9

A polyhedron is a solid formed by plane faces. Is the graph of a polyhedron always planar?

7.4 PATHS, CIRCUITS, AND TREES

In this section we shall introduce some definitions that have proven extremely useful in the study of graphs. The first definition deals with the idea of a path from one node of a graph to another:

Definition

A *path* from node I to J is a sequence of branches such that

    (a) the first branch begins at node I,

    (b) each branch ends at the beginning node of the next,

    (c) the last branch ends at node J,

    (d) no branches are traversed more than once.

This idea arises naturally in the study of communication or transportation networks, for example, where we may be interested in finding paths for messages between communication terminals, or paths for automobiles, trains, or airplanes between cities. Notice that this definition says nothing about the possibility of visiting some intermediate nodes more than once.

As illustrations, the following are all paths from node 1 to 8 in the graph of Fig. 7.2.1:

(1,8)
(1,6),(6,10),(10,9),(9,8)
(1,4),(4,5),(5,9),(9,4),(4,8)

Graphs in which it is possible to find a path between every pair of nodes fall into a special category. Such graphs are in some sense "one-piece" graphs. This motivates the following definition.

Definition

A graph is said to be *connected* if there is at least one path between every pair of nodes.

A graph that is not connected consists of two or more isolated pieces and, for this reason, will usually be of no interest to us. An example of such a graph is shown in Fig. 7.4.1.

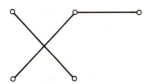

FIG. 7.4.1   A disconnected graph.

Another important idea concerns the possibility of a path returning to its starting node without repeating other nodes:

Definition

A *circuit* is a nonempty path from node *I* to node *I* with the following property: no nodes are visited more than once, except node *I*, which is visited exactly twice—once at the beginning and once at the end.

Referring again to the graph of Fig. 7.2.1, the following are circuits from node 30 to 30:

$$(30,35),(35,28),(28,23),(23,29),(29,30)$$
$$(30,36),(36,31),(31,30)$$

On the other hand, the following are *not* circuits:

$$(30,35),(35,30)$$
$$(30,35),(35,28),(28,29),(29,23),(23,24),(24,29),(29,30)$$

Why not?

Our last definition concerns a class of graphs that plays an important role in all of graph theory:

Definition

A *tree* is a connected graph without circuits.

We have already seen one example of a tree, the natural gas pipeline system shown in Fig. 7.2.3. If a circuit existed in this system, it would be possible for natural gas to circulate without going directly to the collection point. Since no circuits exist, but at the same time the graph is connected, we can guarantee that gas can flow on a unique path from each well to the collection point.

Other important applications of trees arise in problems involving sequences of decisions. Suppose, for example, that a medical doctor is interviewing a prospective patient, and wishes to determine something of his medical history. He might ask, "Do you have any allergies?" If the answer is "No," the doctor would go on to other questions. If the answer is "Yes," the next question might be "Are you allergic to pollen?" If the answer is "No," other allergies would be tried. If "Yes," specific pollens would be suggested. Such a questioning procedure can be represented by a tree, as shown in Fig. 7.4.2.

Only a small part of such a tree is shown. Each node represents a question to be asked at each possible stage, and real medical questionnaires contain thousands of questions. When the end of a line of questioning is reached, it is necessary to return to a higher node in the tree, and to continue to a branch other than the one originally taken from that node.

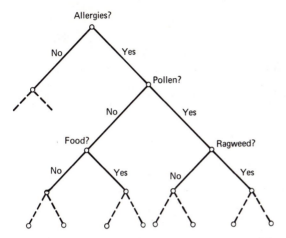

FIG. 7.4.2   A tree representing a medical questionnaire.

Trees enjoy many interesting and useful properties, some of which are described in the next section.

Exercise 7.4.1

Prove that in a tree there is one and only one path between any two nodes.

Exercise 7.4.2

Make up a questionnaire for some purpose that interests you, and draw the corresponding tree.

Exercise 7.4.3

In what sense is a graph theory tree similar to a botanical tree?

Exercise 7.4.4

Let C be a circuit from node I to I that visits nodes $I1, I2, \ldots, IN$ as intermediate nodes. Show that the branches of C also determine a circuit from node $I1$ to $I1$, node $I2$ to $I2$, $\ldots$, and from node IN to IN.

Exercise 7.4.5

We wish to assign each of the integers 1 to 2^n to an n-bit binary number, so that each binary number corresponds to exactly one integer. Call the binary number associated with the integer i, b_i. Consider the problem of making the assignment so that if i and j differ by 1, b_i and b_j differ in only one bit. Interpret this problem as the problem of finding a certain kind of path on the n-cube (see Exercise 7.3.7). Solve the problem for $n = 3$. Can you think of a practical advantage of using such a code to represent samples of a digital signal? (Such codes are called *Gray* codes.)

7.5　SOME PROPERTIES OF TREES

We now shall derive some properties of trees that will be useful to us later, but that also are of general interest.

Property 1

Every tree with at least two nodes has at least one node whose degree is exactly one.

Proof

We shall assume the opposite and arrive at a contradiction, thereby proving the proposition. Assume, then, that there is no node of degree one. Since a tree is connected, every node must therefore have at least two other nodes connected to it. Pick any node, say $I1$, and move to one of the nodes adjacent to it, say $I2$. $I2$ is now adjacent to some other node, say $I3$. Move to $I3$. We may continue in this way, leaving each node we arrive at by a branch we have not yet used. Eventually, we must be forced to return to a node that we have already visited, since there are altogether only a finite number of nodes in the tree. This means we have found a circuit, which is a contradiction.

The nodes of a tree that have degree one are called, naturally enough, *leaves*. If we remove a leaf node and its associated branch, the remaining graph must still be connected, and can have no circuits, since the original tree had no circuits. Therefore, the remaining graph is a smaller tree. We may paraphrase this by saying: picking leaves from a tree still leaves a tree. This fact enables us to establish the next property.

Property 2

Every tree with n nodes has exactly $n-1$ branches.

Proof

We shall prove this by induction on n, the number of nodes in the tree. The property is true for 2 and 3 nodes, as may be verified by drawing all possible graphs for these numbers of nodes. We shall show that if the property is true for $n-1$ nodes, then it must be true for n nodes, and this will establish property 2. Consider a tree with n nodes. Choose a leaf, and remove it, together with its branch. As discussed above, the remaining graph is a tree with $n-1$ nodes. By the induction hypothesis, it has $n-2$ branches. Now adding the leaf back to the tree, we add exactly one branch, making the total $n-1$. This means that the original tree had $n-1$ branches, which is what we wanted to show.

We have shown that a tree, which is a connected graph without circuits, has exactly $n-1$ branches. We can also show that any connected graph with $n-1$ branches cannot have any circuits. Thus, a connected graph with n nodes is a tree *if and only if* it has exactly $n-1$ branches.

Property 3

Every connected graph with n nodes and $n-1$ branches is a tree.

Proof

Suppose the graph has a circuit. Remove any branch on this circuit. The remaining graph is still connected, since any path that used the missing branch can use the remaining part of the circuit. Continue in this way, removing branches on circuits, until the remaining graph has no circuits. This final graph is still connected, has no circuits, and is therefore a tree. But it has fewer than $n-1$ branches, which contradicts property 2.

We have shown in effect that $n-1$ is the minimum number of branches required to bring n nodes together into a connected graph, and that the resulting graph is in fact a tree. If we try to use fewer branches we must have a disconnected graph; while if we use more, we must have a graph with circuits. One consequence of these facts that will prove useful is that if we add a branch to a tree, we always create a circuit.

Exercise 7.5.1

Let $G = (N,B)$ be a connected graph. Prove that we can always find another graph $G' = (N,B')$, such that B' is a subset of B, and such that G' is a tree. (That is, we can always remove branches from a connected graph until we arrive at a tree.)

Exercise 7.5.2

Prove that we can always build any given tree by adding leaves in a certain way to some initial starting node. Discuss an algorithm that will determine the starting node and the order in which the leaves should be added.

Exercise 7.5.3

Prove that every tree is planar.

Exercise 7.5.4

Prove that every tree with at least two nodes has at least two leaves.

7.6 FURTHER EXAMPLES OF TREES

There are many examples of trees provided by logical structures such as the questionnaire tree, and we shall discuss some of those now. These trees all have the property that they start at one node (called the *root*) and develop from there by branching. Consider, for example, the game of chess, where there are always a finite number of alternative moves at each stage of the game. We can represent the starting position by a node, say r. There are 20 possible first moves: 16 Pawn moves and 4 Knight moves. We represent these by 20 nodes at the second level, each connected to the root node r. These 20 nodes represent positions from which Black moves, and these are denoted by $x_1, x_2, \ldots, x_{20}$ in Fig. 7.6.1. Suppose that

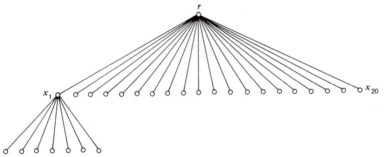

FIG. 7.6.1   A tree for the game of chess.

x_1 represents the move "*P-QR3*". From this position Black will have 20 possible moves, which we represent by 20 nodes at level 3. Since Black has 20 moves for each of the nodes at level 2, there are altogether $20 \times 20 = 400$ nodes at level 3. Then each of the possible White replies is represented by a node, and so forth. If we wish to represent even 10 or 20 moves, the size of the tree becomes too large to draw completely. The use of a tree to represent a game, however, is useful conceptually, since it allows us to think of choosing and evaluating moves as searches ahead on the game tree. At any point in a game, for example, we can construct a tree by first putting in those moves open to us that seem "reasonable." We can then add to the tree nodes that represent "reasonable" replies, and so forth. If we restrict the number of nodes at each level to those representing intuitively sound moves, we can construct a tree that goes quite "deep" along certain lines. In fact, this is believed to approximate what happens in the mind of a good chess player, who is capable of following certain special lines of play many moves into the future.

Another example of a tree is provided by the process of classification. Take for instance the classification of evergreen trees (botanical).* Evergreens are classified first by whether they are erect or creeping. Those that are erect are classified according to whether they have needle-like or scale-like leaves. Those that creep are classified according to whether they are non-flowering or flowering. Figure 7.6.2 shows a small portion of the tree

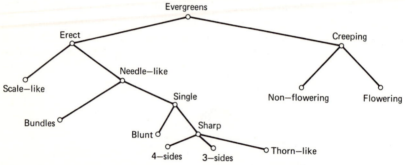

FIG. 7.6.2   Part of a classification tree for evergreen trees.

of evergreens. Such classification schemes are very useful in many fields, especially the empirical biological ones.

As a last example of a tree representing a logical structure, consider the evaluation of the arithmetic expression shown below in Eq. 7.6.1:

$$(x+y)*z+((w*u)+v) \qquad (7.6.1)$$

If this expression is to be evaluated using given values of the variables, we must perform the operations in a certain order. For example, we must add *x* and *y* before multiplying the result by *z*. Figure 7.6.3 shows a tree that represents this arithmetic expression.

The nodes at the bottom level represent the values of the variables x,y,z,u,v,w. Each node at the higher levels represents the value of some expression obtained by performing the indicated operation with the values at the nodes connected to this node at the levels below. For example, the node marked *s* and having a plus sign associated with it means the following: "Assign to node *s* the value obtained by adding the values at the nodes below *s* in the tree: *x* and *y*." Only after the nodes in the tree below

---

* See *A Field Guide to Trees and Shrubs*, G. A. Petrides, Houghton Mifflin, Boston, 1958.

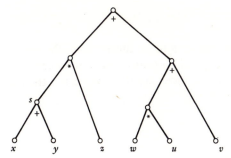

FIG. 7.6.3   A tree that represents an arithmetic expression.

a given node have been assigned values can the value at that given node be assigned a value. The value assigned to the top node (the root) represents the value of the original arithmetic expression. The technique of representing arithmetic expressions by trees is useful in the writing of compilers for producing machine instructions for evaluating expressions, and the process of constructing the tree is called parsing.

Exercise 7.6.1

Consider the class of drawings consisting of circles, some nested within others. An example is shown below.

Describe how such a drawing can be represented by a tree.

Exercise 7.6.2

Is a family tree a tree? Discuss conditions under which it is.

Exercise 7.6.3

Consider the experiment of tossing a coin n times, with the result "heads" or "tails" each time. Show how the result of such an experiment can be represented by a tree. How many nodes are there in such a tree, as a function of n? How many branches?

7.7 CONCLUSION

We have seen that graphs can be used to represent many different practical situations. These situations often give rise to practical questions that are reflected as problems about graphs.

The remainder of this book is devoted to the discussion of four particular such problems: (1) the construction of the shortest tree connecting a set of nodes, (2) the construction of the shortest path connecting two nodes in a graph, (3) finding the maximum amount of some quantity that can be sent between two nodes in a graph with certain capacity limitations, and (4) finding the voltage distribution in a graph representing an electrical network. These problems were chosen to illustrate the development of computer algorithms for solving practical problems associated with graphs.

Further Reading

The following books are devoted solely to graph theory, and will provide the reader with further mathematical development, and more examples of applications. They are listed, roughly, in ascending order of difficulty.

1. *Graphs and Their Uses,* O. Ore, Random House, New York, 1963.

2. *Finite Graphs and Networks,* R. G. Busacker and T. L. Saaty, McGraw-Hill, New York, 1965.

3. *The Theory of Graphs and Its Applications,* C. Berge, Wiley, New York, 1962.

4. *Graph Theory and Finite Combinatorics,* S. S. Anderson, Markham, Chicago, 1970.

5. *Graph Theory,* F. Harary, Addison-Wesley, Reading, Mass. 1969.

6. *Theory of Graphs,* O. Ore, American Mathematical Society, vol. XXXVIII, Colloquium Publications, Providence, R.I., 1962.

A general discussion of the applications of graph theory to the analysis of large systems can be found in reference 7.

7. "Network Analysis," H. Frank and I. T. Frisch, *Scientific American,* vol. 223, no. 1, July 1970, pp. 94–103.

The graph in Fig. 7.2.1 is from reference 8.

8. "The Design of Minimum-Cost Survivable Networks," K. Steiglitz, P. Weiner, and D. J. Kleitman, *IEEE Transactions on Circuit Theory,* vol. CT-16, no. 4, November 1969, pp. 455–460.

8.
THE SHORTEST TREE PROBLEM

8.1 THE PROBLEM

Suppose we want to build a highway network that connects a given set of cities, and we wish to do this with minimum expense. Assuming that the cost of building a highway between any two cities is proportional to the distance between them, and that we are not permitted to introduce junctions except at the cities themselves, we arrive at the following problem in graph theory terms: find a graph that connects a given set of nodes and whose branches have the smallest total length. It is easy to see that such a graph must be a tree, for it must be connected, and if it had a circuit a branch could be removed and the total length reduced. This problem is called the *shortest tree problem.*

As another example of a situation in which this problem arises, consider the construction of a printed circuit board. There may be several electrical terminals, each located at a specific place on the board, all of which need to be connected together electrically. The problem of wiring these together, without introducing other terminals, ignoring other connections on the board, and using the minimum total length of wiring, is another example of the shortest tree problem.

The first idea that might occur to us for solving this problem is enumeration: that is, the process of enumerating every tree that connects a set of nodes, finding the total length of each one, and then choosing the tree with the lowest total. The first stumbling block to carrying out this idea is the problem of writing a computer program that will generate all the trees. We would want, in fact, to generate all the trees without duplicating any, because evaluating the length of a duplicated tree would be a waste of time. This problem is formidable but can be solved, and algorithms for generating trees without duplication have been published in the literature. The next stumbling block is, however, more serious. If the program envisioned, which generated and evaluated trees, were run on a modern computer, for, say 30 nodes, the program would not finish in our lifetimes. To see this, we must investigate the number of trees that connect n nodes. Although deriving this number is somewhat difficult (the actual number of distinct trees is n^{n-2}), we can convince ourselves quite easily that the number is gigantic as follows.

Consider, for example, only those trees with exactly 2 leaves. These trees take the form of a single chain, as shown in Fig. 8.1.1.

FIG. 8.1.1   A tree with exactly two leaves.

There are n choices for the first node, $(n-1)$ for the second, and so forth, so that the total number of chains of the type shown is $n!$. But we have counted each distinct tree twice, since each chain can be formed in two ways, starting from each end. Hence, there are $n!/2$ trees with exactly 2 leaves. This represents a very conservative lower bound on the total number of all trees, but shows immediately that enumeration is out of the question for $n = 30$ or even 20 nodes, since $20! = 2.4 \times 10^{18}$. (The choice of the exclamation point as a symbol for the factorial function is a fortunate one.)

Exercise 8.1.1

Prove that all trees with exactly 2 leaves do, in fact, take the form shown in Fig. 8.1.1.

Exercise 8.1.2

What form does a tree with n nodes and $n-1$ leaves take? How many such distinct trees are there?

8.2 DEVELOPMENT OF AN ALGORITHM

Having encountered the difficulties discussed above with the most straightforward approach to the problem, we may be surprised at how simply the problem lends itself to solution by computer. The algorithm we shall derive is based on certain properties of the shortest tree, which we shall now establish.

Property 1

Let i be any node, and let j be the node closest to i. Then the branch (i,j) is in the shortest tree.

Proof

Suppose that branch (i,j) is not in the shortest tree. Then adding this branch (i,j) to the shortest tree produces a circuit, since nodes i and j were connected by a path before. This situation is shown in Fig. 8.2.1.

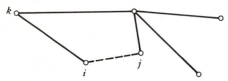

FIG. 8.2.1 The result of adding branch (i,j) to the shortest tree in the proof of Property 1.

As part of this circuit there must be another branch connected to i, say branch (i,k). Branch (i,k) can now be removed, leaving a tree that is shorter than the original, since j is closer to i than is k. This contradiction proves the property.

This property can be generalized as follows.

Property 1'

Let the set of nodes be partitioned into two nonempty sets, V and $\overline{V}$. Let the shortest branch that connects a node in V to a node in $\overline{V}$ be branch (i,j) ; $i\varepsilon V, j\varepsilon \overline{V}$. Then this branch is in the shortest tree.

Proof

To prove this property we proceed in much the same way as before. Suppose that this property does not hold, and that branch (i,j) is not in the shortest tree. Then adding (i,j) to the shortest tree produces a circuit, and in this circuit there is, besides (i,j), some other branch going from a node in V to a node in $\overline{V}$, say branch (k,m). This situation is shown in Fig. 8.2.2.

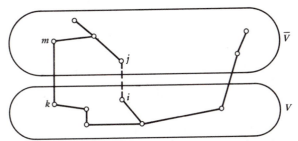

FIG. 8.2.2 The result of adding branch (i,j) to the shortest tree in the proof of Property 1'.

Branch (k,m) can now be removed, leaving a shorter tree, which is a contradiction as before.

This property can be used to devise an algorithm for producing the shortest tree. First we select any node, and put this node, say i, in V. The rest of the nodes comprise the set $\overline{V}$. Now find the shortest branch from a node in V (in this case just node i) to a node in $\overline{V}$ (the rest of the nodes). Call this second node j and add branch (i,j) to our growing tree. Next add node j to V and proceed as before, finding the shortest branch from a node in V to a node in $\overline{V}$. When the last node is connected we have $n - 1$ branches that by Property 1' must be in the shortest tree, and these $n - 1$ branches in fact determine the shortest tree.

Exercise 8.2.1

Devise an algorithm for finding the longest tree connecting a given set of nodes.

8.3 PROGRAMMING THE ALGORITHM

Figure 8.3.1 shows a FORTRAN program that implements the algorithm developed above, and we shall now describe how this program works. First, storage space is allocated to the real, one-dimensional arrays X(I) and Y(I), which will contain the *x*- and *y*-coordinates of node I, respectively. Next, storage space is allocated to the real two-dimensional array DIST(I,J), which will contain the distance between node I and J. This distance matrix will be computed after the coordinates are read in. The tree will be stored in adjacency-list form as it is formed in the program, and the arrays NEAR(I,J) and NUMBER(I) comprise the adjacency-list structure, with NUMBER(I) being the number of nodes adjacent to node I, and NEAR(I,J) being the Jth node adjacent to node I. The program shown provides storage for problems of up to 50 nodes.

The one-dimensional array INV(I) indicates if node I is in the set $\bar{V}$ or not; that is, INV(I) will be 1 if node I is in V, and 0 if not.

The execution of the program begins by reading in N, the number of nodes, and then the coordinates X(I) and Y(I). Next, the arrays NUMBER(I), INV(I), and NEAR(I,J) are all initialized to zero, and the distance matrix DIST(I,J) computed and filled in. After this, N, the coordinates, and the distance matrix are printed out. This brings us down past statement 15.

Next, INV(1) is set equal to 1, indicating that we start with node 1 in the set V. We shall keep track of the number of nodes in V with the variable NODES. When NODES = N we are finished. We initially set NODES = 1.

The code starting with statement 7 and ending with statement 5 finds the shortest branch between a node in V (called ISTAR) and a node in $\bar{V}$ (called JSTAR). After this, the new branch (ISTAR,JSTAR) is put in the tree, by updating NUMBER(ISTAR), NUMBER(JSTAR), NEAR(ISTAR, M), and NEAR(JSTAR,M). The node JSTAR is then put into the set V, by setting INV(JSTAR) equal to 1 and by incrementing NODES by 1. At this point, NODES is tested: if it is less than N we go back to statement 7 and find another branch to add to the tree; if it equals N we are finished, and we go on to print out the branches in the shortest tree. Notice that we print out branch *i,j* only when *j* > *i*; this means that each branch is printed only once.

As an example, the 10 node problem shown in Fig. 8.3.3 was run. The corresponding data cards, with N on the first card and the coordinates on the next 10, are shown on page 204.

```
C......THIS PROGRAM FINDS THE SHORTEST TREE CONNECTING N NODES.
      REAL X(50),Y(50),DIST(50,50)
      INTEGER NEAR(50,50),NUMBER(50),INV(50)
C......READ IN N AND COORDINATES, INITIALIZE ARRAYS
      READ(5,1) N
    1 FORMAT(I2)
      DO 2 I=1,N
    2 READ(5,3)X(I),Y(I)
    3 FORMAT(2F10.5)
      DO 4 I=1,N
      NUMBER(I)=0
      INV(I)=0
      DO 4 J=1,N
      DIST(I,J)=SQRT((X(I)-X(J))**2+(Y(I)-Y(J))**2)
    4 NEAR(I,J)=0
C......WRITE OUT N, COORDINATES, DIST MATRIX
      WRITE(6,11) N
   11 FORMAT('1N=',I5)
      DO 12 K=1,N
   12 WRITE(6,13)X(K),Y(K)
   13 FORMAT(' X=',E14.7,' Y=',E14.7)
      DO 14 K=1,N
   14 WRITE(6,15)(DIST(K,L),L=1,N)
   15 FORMAT(5(' ',F8.3))
C......PUT NODE 1 IN V
      INV(1)=1
      NODES=1
    7 DMIN=1.E10
C......SEARCH OVER ALL PAIRS I IN V AND J IN VBAR FOR SHORTEST DISTANCE
      DO 5 I=1,N
      IF(INV(I).NE.1)GOTO5
C......NODE I IS IN V
      DO 6 J=1,N
      IF(INV(J).EQ.1)GOTO6
C......NODE J IS IN VBAR
      IF(DIST(I,J).GE.DMIN)GOTO6
      ISTAR=I
      JSTAR=J
      DMIN=DIST(I,J)
    6 CONTINUE
    5 CONTINUE
C......PUT IN BRANCH (ISTAR,JSTAR), PUT JSTAR IN V
      M=NUMBER(ISTAR)+1
      NEAR(ISTAR,M)=JSTAR
      NUMBER(ISTAR)=M
      M=NUMBER(JSTAR)+1
      NEAR(JSTAR,M)=ISTAR
      NUMBER(JSTAR)=M
      INV(JSTAR)=1
      NODES=NODES+1
C......GO BACK IF NOT FINISHED
      IF(NODES.LT.N)GOTO7
C......PRINT OUT BRANCH (I,NEAR(I,J)) IF NEAR(I,J) > I
      DO 10 I=1,N
      M=NUMBER(I)
      DO 8 J=1,M
      IF(NEAR(I,J).LE.I)GOTO8
      WRITE(6,9)I,NEAR(I,J)
    8 CONTINUE
   10 CONTINUE
    9 FORMAT(' BRANCH ',I2,' TO ',I2)
      STOP
      END
```

FIG. 8.3.1 A FORTRAN program that finds the shortest tree connecting *N* nodes.

col. 11

10		...N
0.	0.	...X(1),Y(1)
0.	1.	...X(2),Y(2)
0.	3.	
1.	0.	.
1.	2.	.
1.	4.	.
2.	1.	
2.	2.	
2.	3.	
3.	0.	...X(10),Y(10)

$$(8.3.1)$$

Figure 8.3.2 shows the output of the program when it was run on this problem: the number of nodes, N; a list of the coordinates; the DIST matrix; and the branches in the solution. The DIST matrix is printed out in a FORMAT that allows 5 elements per line, so that each row of the 10×10 matrix for this problem takes up 2 lines in the printed output. Finally, Fig. 8.3.3 shows a picture of the shortest tree obtained as a solution.

In the previous discussion we have ignored the possibility that at some stage there may be more than one branch with the same smallest distance from a node in V to a node in $\overline{V}$; that is, that a tie may occur when a search for a new branch is carried out. It turns out that insofar as the total length of the final tree is concerned, it does not matter how ties are broken. In the computer program described above, the first branch encountered with the smallest length is kept. The example above illustrates the fact that the shortest tree may not be unique, since branch (4,7) may be substituted for branch (2,5) without changing the total length.

Example

Another algorithm for finding the shortest tree is described by Kruskal (see reference 2 in the suggestions for further reading at the end of this chapter):

(1) List all possible branches in order of increasing length.

(2) Choose the smallest branch.

```
N=    10
X= 0.0000000E 00  Y= 0.0000000E 00
X= 0.0000000E 00  Y= 0.1000000E 01
X= 0.0000000E 00  Y= 0.3000000E 01
X= 0.1000000E 01  Y= 0.0000000E 00
X= 0.1000000E 01  Y= 0.2000000E 01
X= 0.1000000E 01  Y= 0.4000000E 01
X= 0.2000000E 01  Y= 0.1000000E 01
X= 0.2000000E 01  Y= 0.2000000E 01
X= 0.2000000E 01  Y= 0.3000000E 01
X= 0.3000000E 01  Y= 0.0000000E 00
        0.000      1.000      3.000      1.000      2.236
        4.123      2.236      2.828      3.606      3.000
        1.000      0.000      2.000      1.414      1.414
        3.162      2.000      2.236      2.828      3.162
        3.000      2.000      0.000      3.162      1.414
        1.414      2.828      2.236      2.000      4.243
        1.000      1.414      3.162      0.000      2.000
        4.000      1.414      2.236      3.162      2.000
        2.236      1.414      1.414      2.000      0.000
        2.000      1.414      1.000      1.414      2.828
        4.123      3.162      1.414      4.000      2.000
        0.000      3.162      2.236      1.414      4.472
        2.236      2.000      2.828      1.414      1.414
        3.162      0.000      1.000      2.000      1.414
        2.828      2.236      2.236      2.236      1.000
        2.236      1.000      0.000      1.000      2.236
        3.606      2.828      2.000      3.162      1.414
        1.414      2.000      1.000      0.000      3.162
        3.000      3.162      4.243      2.000      2.828
        4.472      1.414      2.236      3.162      0.000
BRANCH   1  TO   2
BRANCH   1  TO   4
BRANCH   2  TO   5
BRANCH   3  TO   5
BRANCH   3  TO   6
BRANCH   5  TO   8
BRANCH   7  TO   8
BRANCH   7  TO  10
BRANCH   8  TO   9
```

FIG. 8.3.2   Output when the shortest tree program is run on the example.

(3) Among the remaining unchosen branches, choose the shortest that does not form a circuit with the branches already chosen.

(4) Repeat (3) until we have chosen $n-1$ branches. (This is always possible.)

To prove that this procedure produces the shortest tree (assuming no ties in branch length), let the chosen branches be $b_1, \ldots, b_{n-1}$ in the order chosen, let B be the graph with these branches, and let T be the shortest tree.

First, we need to show that B is a tree. That is, any graph with $n-1$ branches and no circuits is connected, and is therefore a tree. Assume the

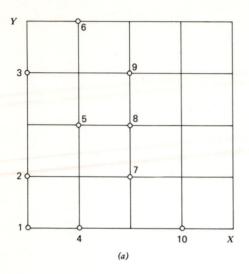

(a)

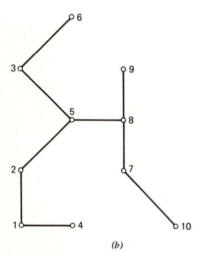

(b)

FIG. 8.3.3 The example for the shortest tree program; (a) coordinates; (b) shortest tree.

opposite: that B consists of $p \geq 2$ disconnected pieces $B_1, \ldots, B_p$; with n_i nodes in each piece, and m_i branches in each piece. It cannot be true that $m_i \leq n_i - 1$ for every i, since this would imply that

$$n - 1 = \sum_{i=1}^{p} m_i \leq \sum_{i=1}^{p} n_i - p = n - p$$

Hence, for some piece B_q, $m_q > n_q - 1$. This piece must have a circuit, which contradicts the way B was constructed.

We now want to show that B is the same tree as T and, hence, that the algorithm above produces the unique shortest tree, assuming that no two branches are equal in length. Suppose not. Let b_k be the first branch in the list $b_1, \ldots, b_{n-1}$ not in T. Add the branch b_k to T, producing a graph T'. This forms a circuit in T' with b_k in the circuit. Not all the branches in the circuit are in B, for otherwise B would have a circuit. Let e be a branch in the circuit not in B. Form the tree T'' by removing e from T'. The graph with branches $b_1, \ldots, b_{k-1}, e$ has all its branches in T and, hence, has no circuits. Hence the length of e is greater than the length of b_k, for otherwise e would have been in the list $b_1, \ldots, b_k$. Therefore T'' is a tree with total length less than T, a contradiction.

Notice that we have obtained as a bonus the result that if no two branches have the same length, the shortest tree is unique. Also, the proof above that any graph with n nodes, $n - 1$ branches, and no circuits is a tree, enables us to state the following:

Theorem

The following three conditions are equivalent for a graph G with n nodes:

    (1) G is connected and has no circuits (G is a tree).

    (2) G has $n - 1$ branches and no circuits.

    (3) G has $n - 1$ branches and is connected.

Proof

    (1) $\Rightarrow$ (2) by property 2 of Section 7.5.

    (2) $\Rightarrow$ (3) by the proof above.

    (3) $\Rightarrow$ (1) by property 3 of Section 7.5.

This theorem can be restated in the following way: of the following three conditions for a graph G with n nodes:

    (a) G is connected.

    (b) G has $n - 1$ branches.

    (c) G has no circuits.

Any two imply the third.

Exercise 8.3.1

Do the example of Fig. 8.3.3 by hand, recording the order in which the branches are added to the tree. Resolve ties the same way that the computer program does, and check your answer with the one given.

Exercise 8.3.2

Suppose the statement in the DO 6 loop

IF(DIST(I,J).GE.DMIN)GOTO6

were changed to

IF(DIST(I,J).GT.DMIN)GOTO6

How would that affect the way in which ties were resolved? What tree of smallest length would result if the modified program were run on the example?

Exercise 8.3.3

How would you modify the program in Fig. 8.3.1 so that the tree found was stored in branch-list form? In adjacency-matrix form?

*Exercise 8.3.4

Prove that the total length of the final tree does not depend on how ties are broken in the algorithm.

Exercise 8.3.5

Discuss the computer implementation of Kruskal's algorithm.

Exercise 8.3.6

(4) *G* is connected, but becomes disconnected if any branch is deleted.

Show that this condition is equivalent to any of the three conditions given in the theorem above. (*Hint:* prove that (4) $\Rightarrow$ (1) by showing that *G* can have no circuits.)

*Exercise 8.3.7

Show that the operation of comparing distances:

IF(DIST(I,J).GE.DMIN)GOTO6

in the algorithm in Fig. 8.3.1 is executed $(N^3 - N)/6$ times. Thus, for large *N*, the program performs approximately $N^3/6$ comparisons. *Hint:* you may need the identities

$$\sum_{i=1}^{N-1} i = \frac{N(N-1)}{2}$$

$$\sum_{i=1}^{N-1} i^2 = \frac{N(N-1)(2N-1)}{6}$$

which can be established by induction.

*Exercise 8.3.8 (computer experiment)

The algorithm of Fig. 8.3.1 is inefficient in the sense that a given branch, if it remains between *V* and $\overline{V}$, will be examined more than once. This may be remedied by saving, for each node in $\overline{V}$, the *shortest distance to a node in V*. When a node *I* is transferred from $\overline{V}$ to *V*, we need then only compare the distances from *I* to the nodes remaining in $\overline{V}$ with the saved information, updating the saved information when necessary. Write a FORTRAN program that implements such an improved algorithm. Show that for large *N*, it requires a number of comparisons of distance proportional to N^2.

8.4 STEINER'S PROBLEM

In the highway network and printed circuit board examples used to introduce the shortest tree problem, we did not allow the introduction of junctions at locations other than the cities themselves. If we do allow such junction points, the problem is called *Steiner's problem,* after J. Steiner, who considered the 3-city problem. The extra nodes that may be introduced are called *Steiner points.*

To see what is involved, consider the problem of connecting 3 cities with a tree of smallest total length, where the cities are unit distances from one another.

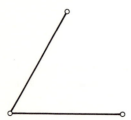

FIG. 8.4.1   A tree connecting three nodes that represent cities.

Shown in Fig. 8.4.1 is the solution to the shortest tree problem, with no Steiner points allowed. The total length is two units. Now consider the tree with a central Steiner point from which radiate three branches to each of the original nodes, as shown in Fig. 8.4.2.

FIG. 8.4.2   A tree that connects the same three cities shown in the previous figure, but that uses an additional point, called a Steiner point.

Each branch has length $1/\sqrt{3}$, so the total length is $\sqrt{3} = 1.732$, which is shorter than the previous tree. Thus, if we wish to build a highway network to connect cities, it is necessary to investigate the possibility of creating intermediate junctions.

Although the 3-city Steiner problem has been solved completely, and much is known about the properties of solutions in general, no practical methods exist for solving problems involving as many as 20 or 30 cities. By a practical method, we mean one capable of being run on present computers in a nonastronomical amount of time. This situation is in sharp contrast with the shortest tree problem, where we developed an algorithm that is quite practical to run on problems involving as many as 50 or 100 cities. Thus, because a problem is easily stated does not imply by any means that it can be solved simply with a computer.

Steiner's problem differs in a fundamental way from the shortest tree problem. In Steiner's problem, there are an *infinite number* of possible solutions to a given problem, whereas in the shortest tree problem there are only a *finite* (although very large) number of possible solutions. Problems with a finite number of possible solutions are called *combinatorial*. Thus, the shortest tree problem is combinatorial, while Steiner's problem is not.

Neither should it be inferred that simple algorithms always exist for solving combinatorial problems. In the next section we shall give an example of a simply stated combinatorial problem which has defied practical solution for many years.

Exercise 8.4.1

Consider the four nodes placed at the vertices of a 2 × 4 unit rectangle. What is a shortest tree connecting the four nodes? Introduce Steiner points so that the shortest tree connecting the augmented set of nodes is shorter than the tree found above.

8.5 THE TRAVELING SALESMAN PROBLEM

A salesman lives in a certain city and plans to make a sales trip to certain other cities. He is to visit each city exactly once and then return home. The problem he faces is the following: in what order should he visit the cities so that the total distance he travels is minimum? This problem, called the *traveling salesman problem*, has been the subject of investigation for more than 30 years; but there is still no simple method of solution.

Techniques capable of solving problems of up to about 40 cities have been developed, with great effort, using today's fastest computers. But the computation time required by these methods for exact solution increases rapidly with the number of cities involved, and it appears that exact solution of problems with 75 or 100 cities will require further theoretical advances.

The problem can be stated in graph theory terms as follows: given N nodes and the distance matrix DIST(I,J), find a circuit that visits every node exactly once before returning to the starting node, and that has the smallest total length. Circuits that touch all the nodes are called *tours*. As an illustration of a traveling salesman problem, we shall go back to a paper published in 1954: "Solution of a Large-Scale Traveling Salesman Problem," by G. Dantzig, D. R. Fulkerson, and S. Johnson, in the *Journal of the Operations Research Society of America* (vol. 2, no. 4, pp. 393–410). In that paper the authors construct a 49-city problem by choosing one city in each state of the continental United States, and the District of Columbia. Road distances between the cities were obtained from an atlas. Figure 8.5.1 shows the geographical location of the 49 cities. Figure 8.5.2 shows the tour that the authors proved optimal; it has a total length of 12,345 miles.

As mentioned in the previous section, the traveling salesman problem

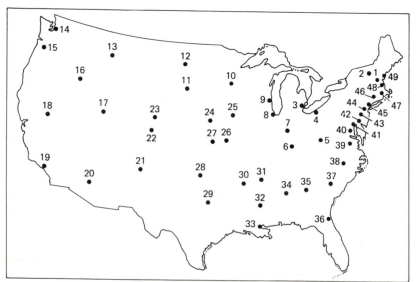

FIG. 8.5.1   A 49-city traveling salesman problem obtained by choosing one city in each continental state, and the District of Columbia.

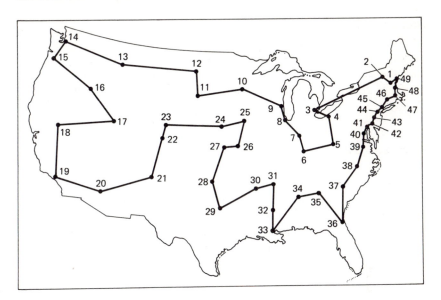

FIG. 8.5.2 A tour for the traveling salesman problem shown in the previous figure. This tour was proved optimal in 1954 by Dantzig, Fulkerson, and Johnson.

is combinatorial; that is, a given problem has only a finite number of possible solutions. We can in fact calculate exactly how many possible solutions there are to an N-city problem as follows. Any tour can be described by a permutation of the node numbers. For example, the tour shown in Fig. 8.5.2 can be described by writing the node numbers visited in order: 17,16,15, . . . ,18. There are altogether $N!$ permutations of N node numbers, so there are altogether $N!$ tours through N cities. But there is duplication among these, because the starting city and the direction are immaterial. Each distinct tour is, in fact, counted $2N$ times. Hence there are $N!/2N = (N - 1)!/2$ distinct tours that represent possible solutions. As in the shortest tree problem, enumeration is not practical except for very small problems.

Recently, computer techniques for approximate solution of large traveling salesman problems have been developed. Such methods appear to provide a promising approach to those problems that do not lend themselves to simple and efficient solution. Just why certain combinatorial problems are so difficult while others are so easy remains somewhat of a mystery. We have no guarantee, in fact, that someone will not invent a very efficient method for exact solution of the traveling salesman problem tomorrow.

Exercise 8.5.1

Describe how a permutation might be stored in a digital computer, and how the cost of its associated tour for the traveling salesman problem would be computed from the distance matrix. Illustrate with FORTRAN.

Exercise 8.5.2

Consider the following algorithm for producing a tour of N cities: start at any city; at each city go to the nearest city that has not yet been visited; return to the starting city when all other cities have been visited. This method is called the *nearest neighbor* method. Give an example showing that it does not always lead to the shortest tour.

Exercise 8.5.3

Consider a 64-node graph, each of whose nodes is associated with a square of a chessboard. Define the distance between any two nodes to be the minimum number of Knight's moves necessary to get between the squares associated with the nodes. The solution to the corresponding traveling salesman problem is a tour with a total distance of 64 moves. What does this mean in terms of Knight's moves on a chessboard?

Exercise 8.5.4

Show that if a traveling salesman problem is defined by a set of cities all on the circumference of a circle, that the solution is given by visiting the cities in the order in which they occur on the circumference.

Exercise 8.5.5 (computer experiment)

Write a computer program that determines the nearest neighbor solution to a traveling salesman problem. The input data should be:

(1) N, the number of nodes;

(2) NSTART, the starting city for the nearest neighbor method;

(3) X(I), Y(I),I = 1,N, the coordinates of the cities in the plane.
Test your program with some actual data obtained from a real source, such as a campus map, a road map, or an airline map. Can you find a better tour by eye? Try running the program with the same data but with several different starting cities. Does the cost of the solution depend much on the starting city?

*Exercise 8.5.6 (computer experiment)

Shen Lin of Bell Telephone Laboratories studied a method for obtaining approximate solutions to the traveling salesman problem called *2-opt*. The method is based on the following ideas. A *2-change* on a tour is a change brought about by breaking two branches in the tour and recombining the remaining two pieces with two other branches to form a new tour. A 2-change is illustrated below in Fig. 8.5.3:

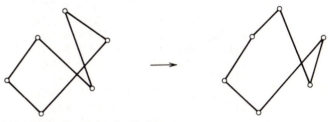

FIG. 8.5.3 Example of a 2-change.

If a 2-change results in a reduction in the total distance of the tour, it is called *favorable*. The 2-opt method consists of the following algorithm:

(1) Start with any tour (possibly a randomly generated one).

(2) Generate all possible 2-changes, one at a time.

(3) If a favorable 2-change is found, adopt the changed tour and go back to 2; if none is found go to 4.

(4) Stop; a tour has been found which has no favorable 2-changes.

Thus, the method consists of searching for successive improvements until no more can be found.

Program 2-opt and run your program on some real examples, as in Exercise 8.5.5. For each problem, try several different starting tours. If you have a random number generator available, try generating random starting tours.

Exercise 8.5.7

R. Mott has suggested the following source of data for a traveling salesman problem.* Once upon a time there was a ring salesman named Frodo from Middle Earth. Below are the coordinates of his sales stops, which he must visit from his home in Hobbiton: what order should he choose for his itinerary?

FORLOND	13 MILES EAST BY 80 MILES NORTH	
HARLINDON	16 MILES EAST BY 64 MILES NORTH	
HOBBITON	28 MILES EAST BY 77 MILES NORTH	
EVENDIM L.	30 MILES EAST BY 85 MILES NORTH	
FORNOST	37 MILES EAST BY 84 MILES NORTH	
THARBAD	42 MILES EAST BY 62 MILES NORTH	
WEATHERTOP	45 MILES EAST BY 78 MILES NORTH	
ANFALAS	46 MILES EAST BY 31 MILES NORTH	
EREGION	55 MILES EAST BY 68 MILES NORTH	
ISENGARD	56 MILES EAST BY 51 MILES NORTH	
RIVENDELL	57 MILES EAST BY 77 MILES NORTH	
EDORAS	58 MILES EAST BY 45 MILES NORTH	
BELFALAS	59 MILES EAST BY 26 MILES NORTH	
DOL AMROTH	62 MILES EAST BY 31 MILES NORTH	
GUNDABAD	63 MILES EAST BY 91 MILES NORTH	
CORSAIRS	65 MILES EAST BY 6 MILES NORTH	
LORIEN	66 MILES EAST BY 60 MILES NORTH	
ENTWASH	67 MILES EAST BY 44 MILES NORTH	
CELEBRANT	69 MILES EAST BY 57 MILES NORTH	
DOL GULDUR	72 MILES EAST BY 63 MILES NORTH	
RAUROS	73 MILES EAST BY 44 MILES NORTH	
MI. TIRITH	80 MILES EAST BY 36 MILES NORTH	
ESGAROTH	83 MILES EAST BY 81 MILES NORTH	
EREBOR	84 MILES EAST BY 86 MILES NORTH	
UDUN	85 MILES EAST BY 43 MILES NORTH	
DAGORLAD	89 MILES EAST BY 50 MILES NORTH	
BARAD DUR	92 MILES EAST BY 41 MILES NORTH	
NURN	95 MILES EAST BY 30 MILES NORTH	
RHUN SEA	109 MILES EAST BY 65 MILES NORTH	

* See *The Lord of the Rings* (parts I to III), J. R. R. Tolkien, Houghton Mifflin, Boston, 1967.

Further Reading

The shortest tree algorithm is due to Prim and is described in reference 1.

1. "Shortest Connection Networks and Some Generalizations," *Bell System Technical Journal*, R. C. Prim, vol. 36, 1957, pp. 1389–1402.

Kruskal's algorithm is described in the following reference

2. "On the Shortest Spanning Subtree of a Graph and the Traveling Salesman Problem," J. B. Kruskal Jr., *Proc. American Mathematical Society*, vol. 7, Feb. 1956, pp. 48–50.

Steiner's problem comes up in the discussion of soap films in reference 3.

3. *What is Mathematics?*, R. Courant and H. Robbins, Oxford University Press, New York, 1951, pp. 354–361.

The following paper treats the problem in detail and refers to many related works.

4. "Steiner Minimal Trees," E. N. Gilbert and H. O. Pollak, *SIAM Journal of Applied Math.*, vol. 16, no. 1, 1968, pp. 1–29.

The traveling salesman problem is surveyed in reference 5.

5. "The Traveling Salesman Problem: A Survey," M. Bellmore and G. L. Nemhauser, *Operations Research*, vol. 16, no. 3, May–June 1968, pp. 538–558.

A traveling salesman problem composed of 49 cities, one in each of the 48 continental states and Washington, D.C., is solved in reference 6.

6. "Solution of a Large-Scale Traveling Salesman Problem," G. Dantzig, D. R. Fulkerson, and S. Johnson, *Operations Research*, vol. 2, no. 4, 1954, pp. 393–410.

A heuristic, approximate approach is described in

7. "Computer Solutions to the Traveling Salesman Problem," S. Lin, *Bell System Technical Journal*, vol. 44, 1965, pp. 2245–2269.

9.
PATHFINDING
ALGORITHMS

9.1 THE SHORTEST PATH PROBLEM

We now take up a problem that is reminiscent of the shortest tree problem, since the criterion is to minimize the total length of some branches, but that requires some new ideas for its solution. Suppose we are given a graph in which each branch has associated with it a distance (or, in more general terms, a cost for traversing it, such as time or money). If we find ourselves at one node and wish to travel along branches to another node, we very naturally may want to find the path with the smallest total length for accomplishing this. This problem was one of the first in graph theory to be solved by computer, and is called the *shortest path problem*. This chapter is devoted to an algorithm for solving this problem which is particularly well suited to the digital computer, and which illustrates the idea of *labeling*.

As in the shortest tree problem, the information about the lengths of the branches will be stored in a two-dimensional array called DIST(I,J), which contains in its (I,J)th place the length* of branch (I,J). For the purposes of this chapter, we shall take DIST(I,J) to be an INTEGER array. Of course, some of the possible branches may not be present in our graph. Since we shall have a description of which branches are in the graph, such as one of the three storage forms for graphs described in Chapter 7, the lengths assigned to the nonexistent branches are irrelevant. Also irrelevant are the numbers on the diagonal, DIST(I,I). Notice also that the distance from I to J is the same as the distance from J to I, so that DIST(I,J) = DIST(J,I) and the array DIST(I,J) will be symmetric about its diagonal.

---

* It is assumed in what follows that no distances are negative. See Exercise 9.3.4 and reference 1 in the suggestions for further reading at the end of this chapter.

As an example, consider the graph shown in Fig. 9.1.1, with the branch lengths indicated by numbers enclosed in circles. This graph might represent a map of airline flights, and the lengths of the branches might represent the air distance between the cities, measured, say, in hours.

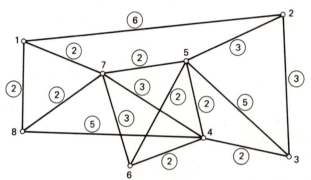

FIG. 9.1.1 Graph used to illustrate the shortest path problem. Branch lengths are indicated by circled numbers.

The distance array of this graph is

I	J= 1	2	3	4	5	6	7	8
1	—	6	—	—	—	—	2	2
2	6	—	3	—	3	—	—	—
3	—	3	—	2	5	—	—	—
4	—	—	2	—	2	2	3	5
5	—	3	5	2	—	2	2	—
6	—	—	—	2	2	—	3	—
7	2	—	—	3	2	3	—	2
8	2	—	—	5	—	—	2	—

$$(9.1.1)$$

Other storage arrangements for the length information are possible. For example, one could use an array LENGTH(I,J) corresponding to the NEAR(I,J) array used in the adjacency-list method of storing graphs. Here, we would store in LENGTH(I,J) the length of the branch (I,NEAR(I,J)).

The arrays NUMBER(I), NEAR(I,J), and LENGTH(I,J) for the same graph as above are shown below:

I	NUMBER(I)	NEAR(I,J),J= 1	2	3	4	5	6	7	8
1	3	2	7	8	—	—	—	—	—
2	3	1	5	3	—	—	—	—	—
3	3	2	5	4	—	—	—	—	—
4	5	3	5	6	7	8	—	—	—
5	5	2	7	6	4	3	—	—	—
6	3	5	7	4	—	—	—	—	—
7	5	1	5	6	4	8	—	—	—
8	3	1	7	4	—	—	—	—	—

$$(9.1.2)$$

I	LENGTH(I,J),J= 1	2	3	4	5	6	7	8
1	6	2	2	—	—	—	—	—
2	6	3	3	—	—	—	—	—
3	3	5	2	—	—	—	—	—
4	2	2	2	3	5	—	—	—
5	3	2	2	2	5	—	—	—
6	2	3	2	—	—	—	—	—
7	2	2	3	3	2	—	—	—
8	2	2	5	—	—	—	—	—

$$(9.1.3)$$

A dash has been used above to signify that the number at that place in the array is irrelevant in the sense that we shall never have reason to use it in a computer program. We sometimes refer to such entries as "don't cares."

Exercise 9.1.1

Give a method for storing in one two-dimensional array the information as to which branches are in the graph and, at the same time, the length of each branch.

Exercise 9.1.2

Discuss the relative merits of the two methods described above for storing the distance information for a graph.

Exercise 9.1.3

What are the storage requirements of each of the two methods?

Exercise 9.1.4

Since the DIST(I,J) array is symmetric and the diagonal elements are meaningless, it is not necessary to take up N^2 storage words (where N is the number of nodes) by storing the full two-dimensional array. Describe a method for storing all possible information in the DIST(I,J) array in a one-dimensional array of length $N(N-1)/2$.

Exercise 9.1.5

Suppose we know that only a small number of branches are actually present in a graph. Devise a method for storing the lengths of these branches that uses less storage space than the two methods described above. Discuss any relative disadvantages of your method.

9.2 A LABELING ALGORITHM

We shall now describe an algorithm for solving the shortest path problem. This algorithm is one of a general class called *labeling algorithms,* so named because at each step certain nodes are assigned *labels* that contain information used to find the solution. The labels for this problem will consist of two parts, which we shall call LAB1(I) and LAB2(I). LAB1(I) will contain the number of the node from which we arrived at node I; LAB2(I) will contain the total distance traversed to get to node I from the starting node K.

The algorithm can be visualized as follows: imagine that we send out messengers from node K along all possible branches. We watch all messengers and wait for the first one to arrive at a new node. When a messenger arrives at a new node, say I, we label that node as follows: we set LAB1(I) equal to K to indicate that this node was reached from node K; we set LAB2(I) equal to the distance traversed in traveling from node K to I. We then instantaneously dispatch new messengers from node I along all possible branches from there. There are now messengers traveling away from nodes K and I. The next messenger to reach a new (unlabeled) node is noted, the new node labeled accordingly, and the process continued, until the destination node L is reached. Whenever a new node is reached, say node J, the second part of its label, LAB2(J), is set equal to the total distance traveled thus far in reaching it. This is obtained by adding the LAB2 label of the preceding node to the length of the branch just traversed.

Of course, we have assumed in our visualization that all messengers travel at the same speed. Hence, each time a LAB2 label is assigned, its value, which is the total distance traveled to date, is greater than all previous ones. For this reason, we need never worry about changing a label at a node once it is assigned, for we can never shorten the path used to reach that node. When we reach the destination node L, we have therefore found the shortest (quickest) path to it from K. The actual path can be reconstructed by backtracking through the nodes given in the LAB1 part of the labels.

To illustrate this algorithm, let us find the shortest path from node 1 to 3 in the same airline graph considered above. We shall indicate a label at a node by [a,b], where a is the LAB1 part and b is the LAB2 part. We have also redrawn the graph so that no branches cross. Starting from node 1, we send three messengers out towards nodes 2,7, and 8. Nodes 7 and 8 are reached simultaneously and it does not matter which we label first; once one is labeled the other can be labeled immediately thereafter. This leads to the graph with three labels shown in Fig. 9.2.1, the label on the starting node being taken as [−1,0] so we know it is labeled.

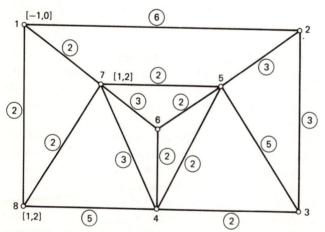

FIG. 9.2.1   The result of labeling nodes 7 and 8 in the example.

To find out the next node to be labeled, we must find out which node is next reached with the smallest total distance. This can be done by examining each labeled node and adding to its **LAB2** the length of each branch that leads from it to an unlabeled node. The result will be placed in the **LAB2** of the next node to be labeled. If we do this for the graph above, we find that node 5 is reached from node 7 with a total length of 4. Next, in order, the following labels are assigned: nodes 6 and 4 are reached from node 7 with total length 5; node 2 from node 1 with length 6; and finally node 3 is reached from node 4 with total length 7. The final labeled graph is shown in Fig. 9.2.2. The shortest path can be traced back from node 3 using **LAB1** at successive nodes, with the resulting path $(3,4),(4,7),(7,1)$.

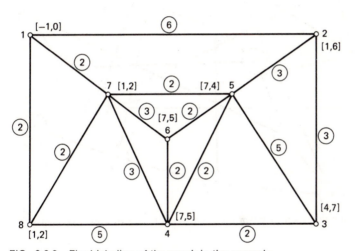

FIG. 9.2.2   Final labeling of the graph in the example.

Notice that when we are done we are left with some bonus information. Every label gives enough information to find the shortest path from node 1 to the labeled node.

Exercise 9.2.1

Make up a shortest path problem and solve it by hand using the labeling algorithm described above.

Exercise 9.2.2

Find the path from *A* to *B* in the graph shown in Fig. 9.2.3 with the fewest number of branches. (This problem was conceived by E. F. Moore.*)

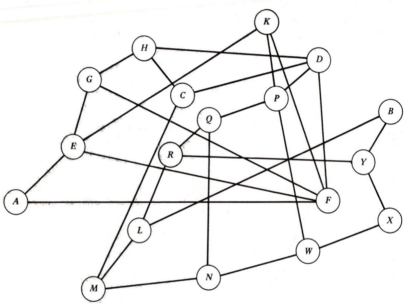

FIG. 9.2.3 A shortest path problem invented by E. F. Moore (see Exercise 9.2.2).

Exercise 9.2.3

Figure 9.2.4 shows a map with road distances indicated in minutes on each branch. Find the quickest route from Philadelphia to New York using the labeling algorithm.

---

* See "The Shortest Path Through a Maze," Edward F. Moore, *Proceedings of an International Symposium on the Theory of Switching,* Part II, The Computation Laboratory of Harvard University *Annals 30,* 1959; pp. 285–292.

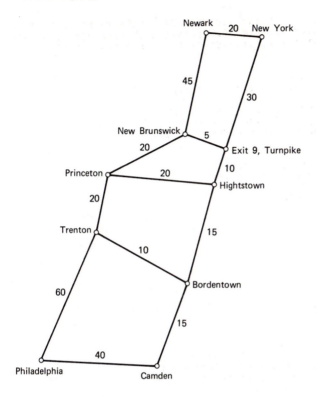

FIG. 9.2.4   Shortest path problem for Exercise 9.2.3.

9.3 PROGRAMMING THE SHORTEST PATH LABELING ALGORITHM

The algorithm described above is ideally suited for digital computation. A complete program is shown in Fig. 9.3.1, and will be described in this section. The program begins by reading in enough information to set up the graph description and the distance matrix. The adjacency-list method is used, with the degree array NUMBER(I) and the adjacency-list NEAR (I,J). The elements of the DIST(I,J) matrix are read in from data cards, and the "don't care" elements will usually be left blank, and so will be read in as 0. Also, the first card contains N, the number of nodes, and K and L, the origin and destination nodes, respectively. When this information is read in, it is also printed out to provide a check of the input data and a record of the problem. For the airline example above, the input data cards are shown on page 227.

```
C......THIS PROGRAM FINDS THE SHORTEST PATH BETWEEN NODES K AND L
       INTEGER NUMBER(50),NEAR(50,50),DIST(50,50),LAB1(50),LAB2(50)
C......READ AND PRINT N,K,L,ADJACENCY-LIST, AND DIST MATRIX
       READ(5,1)N,K,L
     1 FORMAT(50I2)
       WRITE(6,10)N,K,L
    10 FORMAT('1N=',I3,' K=',I3,' L=',I3)
       DO 2 I=1,N
       READ(5,1)M,(NEAR(I,J),J=1,M)
       WRITE(6,11)I,(NEAR(I,J),J=1,M)
    11 FORMAT(' NODE ',I3,' CONNECTED TO',25(I3))
     2 NUMBER(I)=M
       DO 3 I=1,N
       READ(5,1)(DIST(I,J),J=1,N)
     3 WRITE(6,12)(DIST(I,J),J=1,N)
    12 FORMAT(' ',25(I3))
C......INITIALIZE LABELS
       DO 4 I=1,N
       LAB1(I)=0
     4 LAB2(I)=0
       LAB1(K)=-1
     7 LSTAR=100000
       DO 5 I=1,N
       IF(LAB1(I).EQ.0)GOTO5
C......I IS A LABELED NODE
       M=NUMBER(I)
       IF(M.EQ.0)GOTO5
       DO 6 J=1,M
       JI=NEAR(I,J)
       IF(LAB1(JI).NE.0)GOTO6
C......JI IS AN UNLABELED NODE NEXT TO I
       IF(LAB2(I)+DIST(I,JI).GE.LSTAR)GOTO6
C......WE HAVE FOUND A SMALLER TOTAL LENGTH
       LSTAR=LAB2(I)+DIST(I,JI)
       ISTAR=I
       JSTAR=JI
     6 CONTINUE
     5 CONTINUE
C......LABEL JSTAR
       LAB1(JSTAR)=ISTAR
       LAB2(JSTAR)=LSTAR
C......FIND OUT IF WE ARE DONE
       IF(LAB1(L).EQ.0)GOTO7
C......WE ARE DONE
       LAST=L
     9 LAST=LAB1(LAST)
       WRITE(6,8)LAST
     8 FORMAT(' NODE ',I2)
       IF(LAST.EQ.K)STOP
       GOTO9
       END
```

FIG. 9.3.1   A FORTRAN program for finding the shortest path between two nodes of a graph.

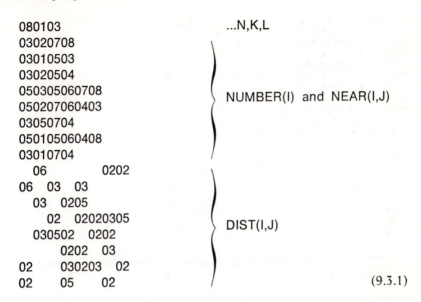

```
080103                          ...N,K,L
03020708
03010503
03020504
050305060708                    NUMBER(I) and NEAR(I,J)
050207060403
03050704
050105060408
03010704
   06          0202
 06  03  03
    03  0205
       02  02020305           DIST(I,J)
    030502  0202
       0202  03
 02      030203  02
 02      05      02                        (9.3.1)
```

Next, the DO 4 loop sets all the labels equal to 0, and then the LAB1 of node K is set to -1, the first label.

Statement 7 begins the main part of the algorithm, which consists of finding for each labeled node the unlabeled nodes that can be reached, and their total distance. The variable LSTAR stands for the new minimum total distance, ISTAR for the new LAB1 label, and JSTAR for the new labeled node. The DO 5 loop finds this new minimum total distance, and we come past statement 5 with values assigned to LSTAR, ISTAR, and JSTAR. Next, node JSTAR is given the appropriate labels, and the LAB1 of L is tested to see if we are done. If the LAB1 of node L is zero we are not done, and so we return to statement 7 for a new stage of labeling.

When the destination node L is finally reached, we backtrack to find the shortest path, printing each node as we go. Figure 9.3.2 shows the output when the airline example is run.

We have assumed that there is some path from K to L. If K and L were not connected, we would eventually come out of the loop ending in statement 5 with LSTAR = 100000, its initial value. We could test for this condition and stop if it occurs.

```
N=  8 K=  1 L=  3
NODE    1 CONNECTED TO  2  7  8
NODE    2 CONNECTED TO  1  5  3
NODE    3 CONNECTED TO  2  5  4
NODE    4 CONNECTED TO  3  5  6  7  8
NODE    5 CONNECTED TO  2  7  6  4  3
NODE    6 CONNECTED TO  5  7  4
NODE    7 CONNECTED TO  1  5  6  4  8
NODE    8 CONNECTED TO  1  7  4
   0  6  0  0  0  0  2  2
   6  0  3  0  3  0  0  0
   0  3  0  2  5  0  0  0
   0  0  2  0  2  2  3  5
   0  3  5  2  0  2  2  0
   0  0  0  2  2  0  3  0
   2  0  0  3  2  3  0  2
   2  0  0  5  0  0  2  0
NODE    4
NODE    7
NODE    1
```

FIG. 9.3.2 The output of the program shown in the previous figure when run on the airline example.

Exercise 9.3.1

Explain what ultimately happens in the execution of the program in Fig. 9.3.1 if there is no path from node K to L. Add FORTRAN code to stop if this situation is encountered.

Exercise 9.3.2

Suggest a way to avoid the necessity of supplying NUMBER(I) at the beginning of each data card that contains a row of NEAR(I,J).

Exercise 9.3.3

Suppose we add a constant x to every element of the distance matrix DIST(I,J). Is the shortest path between given nodes changed thereby?

*Exercise 9.3.4

Consider a shortest path problem for a network in which some of the elements of the distance matrix DIST(I,J) are negative. Will the labeling algorithm work?

9.4 EFFICIENCY OF THE ALGORITHM

The algorithm described above is very efficient compared with other methods which might occur to us. At each labeling step, less than N lengths need to be computed from each of the labeled nodes; and there are less than N of these. The process terminates after a maximum of $(N - 1)$ labeling steps, so altogether less than N^3 additions must be performed. In a 30 node problem, $N^3 = 27000$, which at one addition per microsecond requires only .027 seconds of computer time.

In contrast, consider the computation involved if enumeration were used; that is, if every possible path were evaluated to find the shortest one. In an N node complete graph, there is 1 path between a given pair of nodes with no intermediate nodes visited; $(N - 2)$ paths with 1 intermediate node; $(N - 2)(N - 3)$ paths with 2 intermediate nodes; and so on, up to $(N - 2)!$ paths with $(N - 2)$ intermediate nodes. (We have ignored paths that repeat nodes.) Therefore, there are more than $(N-2)!$ paths that must be evaluated, each requiring more than one addition. Thus, for a 30-node graph, we need to perform more than $28! = 3 \times 10^{29}$ additions; which at one addition per microsecond requires about 10^{16} years, which is about a million times the life of the universe as estimated by current cosmological theory! (An observation of Dr. H. Frank.)

With respect to the existence of very efficient solutions the shortest path problem is similar to the shortest tree problem. Both problems are practically impossible to solve by enumeration, but can be solved quite readily by rather simple algorithms.

*Exercise 9.4.1

Show that the operation of comparing distances:

IF(LAB2(I) + DIST(I,JI).GE.LSTAR)GOTO6

in the algorithm in Fig. 9.3.1 is executed at most $(N^3 - N)/6$ times. (See Exercise 8.3.7.)

*Exercise 9.4.2 (computer experiment)

The algorithm of Fig. 9.3.1 can be improved in much the same way as the shortest tree algorithm discussed in Section 8.3; that is, by saving in temporary labels the *shortest distance through permanently labeled nodes to the starting node.* Write a FORTRAN program that implements such an improved algorithm and show that for large N it requires a number of comparisons of distance proportional to N^2.

9.5 THE PROBLEM OF FINDING ANY PATH

In some situations it suffices to find any path between a given pair of nodes; it is not important to find the shortest such path. As an example, consider the problem of getting out of the maze shown in Fig. 9.5.1.

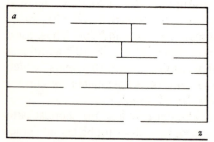

FIG. 9.5.1   A maze.

Suppose we start at point a and we wish to exit at point z. Assign to each passageway a letter, and draw a graph with nodes corresponding to these passageways. Connect two nodes with a branch if their respective passageways connect to each other. Node assignments for the graph above are shown in Fig. 9.5.2, and the corresponding graph is shown in Fig. 9.5.3.

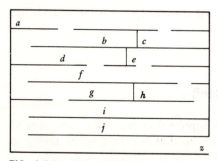

FIG. 9.5.2   Node assignments for the maze.

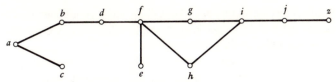

FIG. 9.5.3   Graph representing the maze.

Any path connecting a with z gives a solution to our problem. Of course this maze is simple enough to be solved easily by inspection. A more complicated maze with many more passageways and dead ends might very well require a computer for solution.

Another example of this problem might occur in a telephone switching network, where we are required to find some connection when a call is placed, but there is no particular need for a very efficient path. If such a procedure were automated, we would need a computer algorithm to solve the problem without human intervention.

One way to attack the problem of finding any path is simply to consider it as a special case of the shortest path problem. We can assign arbitrary length to each branch; then the shortest path will certainly be a path and satisfy our requirements. However, the shortest path algorithm is more complicated and less efficient than necessary for this application. We shall now discuss a simplified labeling algorithm that finds some path between a given pair of nodes; we shall call this the *any-path algorithm* to distinguish it from the shortest path algorithm.

To begin with we need only one label at each node, called simply LABEL. LABEL(I) will contain the number of the node from which we arrived at node I (this performs the same function that LAB1 did before). At each labeling step we shall not bother to find the unlabeled node with the smallest total distance; rather, we shall label *all* unlabeled nodes that can be reached from a labeled node. In this way many nodes may be labeled at each step.

When we label away from a node, every adjacent node not already labeled becomes labeled. Therefore, once we label from a node, we need no longer try to label from it again. We shall call a node "scanned" once we have labeled from it. The candidates for nodes to label from are then those that are *labeled* but *unscanned*. Somehow, we must keep enough information in our algorithm to find labeled, unscanned nodes to scan. One way is to keep a list containing the state of each node: unlabeled and unscanned, labeled and unscanned, or labeled and scanned. We can then search the state-list for a node in the second state. We shall, in fact, make use of this method later on, in Chapter 10. At this point it is instructive to use another method of finding labeled, unscanned nodes; one that makes use of the idea of a *stack*.

A stack is simply a list of items, say STACK(I), that is managed in a certain way. Besides the list itself, we keep a single integer variable called POINT, which always tells us how many items are on the stack. We visualize the stack as a vertical pile of items, with only the topmost item accessible. Figure 9.5.4 illustrates a stack with 4 items on it, and shows why we use the name "POINT"; this variable simply "points" to the top of the stack.

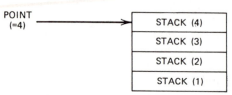

FIG. 9.5.4   Stack and pointer.

When a previously unlabeled node, say J, becomes labeled, it is put on the stack by incrementing POINT and storing J at the new location:

$$\text{POINT} = \text{POINT} + 1$$
$$\text{STACK(POINT)} = \text{J}$$

Similarly, when we need a new node I to scan, it is retrieved from the stack, and POINT is decreased by one:

$$\text{I} = \text{STACK(POINT)}$$
$$\text{POINT} = \text{POINT} - 1$$

Figure 9.5.5 shows a complete FORTRAN program for the any-path labeling algorithm. The program starts as before, with data input and output. Since we are not interested in the lengths of paths, only the adjacency-list information is required. We start with a label of "−1" on node K, the starting node; and only K on STACK. The main loop begins by testing if POINT is zero. If this is the case, we have scanned every labeled node without reaching the destination node L. Therefore there is no path from K to L, and we jump to statement 99 where we print out an appropriate message and stop. If POINT is greater than zero, we take a labeled, un-scanned node off STACK, and scan it by labeling all unlabeled adjacent nodes. Newly labeled nodes get put on STACK. After scanning a node, we test LABEL(L), and go back to the beginning of the main loop, statement 6, if node L has not yet been labeled. If L has been reached, we backtrack and print out the nodes on the path in reverse order, as before.

```
C......THIS PROGRAM FINDS ANY PATH BETWEEN NODES K AND L
      INTEGER NUMBER(50),NEAR(50,50),LABEL(50),STACK(50),POINT
C......READ AND PRINT N,K,L, AND ADJACENCY-LIST
      READ(5,1) N,K,L
    1 FORMAT(50I2)
      WRITE(6,2) N,K,L
    2 FORMAT('1N=',I3,' K=',I3,' L=',I3)
      DO 4 I=1,N
      READ(5,1) M,(NEAR(I,J),J=1,M)
      WRITE(6,3) I,(NEAR(I,J),J=1,M)
    3 FORMAT(' NODE ',I3,' CONNECTED TO',25(I3))
    4 NUMBER(I)=M
C......INITIALIZE STACK AND LABELS
      DO 5 I=1,N
      STACK(I)=0
    5 LABEL(I)=0
C.....PUT K ON STACK AND LABEL IT
      STACK(1)=K
      POINT=1
      LABEL(K)=-1
C......BEGINNING OF MAIN LOOP
C......IS THE STACK EMPTY? IF YES, PRINT MESSAGE
    6 IF(POINT.EQ.0)GOTO99
C......TAKE A LABELED, UNSCANNED NODE, I, OFF THE STACK
      I=STACK(POINT)
      POINT=POINT-1
C......SCAN I
      M=NUMBER(I)
      DO 7 J=1,M
      JI=NEAR(I,J)
      IF(LABEL(JI).NE.0) GOTO7
C......JI IS AN UNLABELED NODE NEXT TO I, LABEL IT AND PUT IT ON STACK
      LABEL(JI)=I
      POINT=POINT+1
      STACK(POINT)=JI
    7 CONTINUE
C......ARE WE FINISHED? IF NO, GO BACK TO BEGINNING OF MAIN LOOP
      IF(LABEL(L).EQ.0)GOTO6
C......BACKTRACK
      LAST=L
    8 LAST=LABEL(LAST)
      WRITE(6,9)LAST
    9 FORMAT(' NODE ',I2)
      IF(LAST.EQ.K)STOP
      GOTO8
   99 WRITE(6,10)
   10 FORMAT(' NODES ARE NOT CONNECTED')
      STOP
      END
```

FIG. 9.5.5   A FORTRAN program for finding any path between two nodes of a graph.

Figure 9.5.6 shows the output when the program was run on the airline graph that was used as an example for the shortest path problem. The path found is (3,4), (4,8), (8,1), which is different from the shortest path.

```
N=   8 K=   1 L=   3
NODE    1 CONNECTED TO  2  7  8
NODE    2 CONNECTED TO  1  5  3
NODE    3 CONNECTED TO  2  5  4
NODE    4 CONNECTED TO  3  5  6  7  8
NODE    5 CONNECTED TO  2  7  6  4  3
NODE    6 CONNECTED TO  5  7  4
NODE    7 CONNECTED TO  1  5  6  4  8
NODE    8 CONNECTED TO  1  7  4
NODE    2
NODE    1
```

FIG. 9.5.6   The output of the program shown in the previous figure when run on the airline example.

Exercise 9.5.1

Show that the any-path algorithm tests labels at most $2B$ times, where B is the number of branches; and hence that the number of label tests is at most $N(N-1)$.

Exercise 9.5.2

If a list is managed as a stack, the most recently added item is used first. This is sometimes referred to as a "last-in-first-out" (LIFO) algorithm. How would you manage a list so that the oldest item was used first? This is called "first-in-first-out" (FIFO), and corresponds to a queue (such as might be found at a theatre).

Exercise 9.5.3

Show that if the list of labeled, unscanned nodes is managed on a first-in-first-out basis, that the any-path algorithm finds a path with the fewest branches.

Exercise 9.5.4

Consider a tree with N nodes. The branch (i_1,i_2), which was not in the tree, is added. This forms a circuit with branches $(i_1,i_2),(i_2,i_3),\ldots,(i_k,i_1)$. Describe an algorithm for finding the intermediate nodes $i_3,\ldots,i_k$ in the circuit.

Exercise 9.5.5 (computer experiment)

The nodes in any graph G can be partitioned into sets $S_1,\ldots,S_p$; where all the nodes in any set are connected by paths to one another; and any node in one set is not connected by a path to any node in any other set. These sets of nodes are called the *components* of the graph G. Write a computer program that determines the components of a graph G. The input data should be:

(1) N, the number of nodes.

(2) M, the number of branches.

(3) NODE1 (I), NODE2 (I), I = 1, . . . ,M; the nodes associated with the M branches.

The output should give the number of components and a list of the nodes in each component. Test your program on a graph with several components.

Further Reading

Pathfinding algorithms are discussed in many of the general references on graph theory mentioned at the end of Chapter 7. In addition, the following paper surveys the literature and will provide the reader with many references.

1. "An Appraisal of Some Shortest-Path Algorithms," S. E. Dreyfus, *Operations Research,* vol. 17, 1969, pp. 395–412.

Dreyfus discusses in detail the algorithm suggested in Exercise 9.4.2, and attributes it to Dijkstra.

2. "A Note on Two Problems in Connexion with Graphs," E. W. Dijkstra, *Numerische Mathematik* 1, 1959, pp. 269–271.

In addition, he discusses the shortest path problem when some distances are allowed to be negative, the problem of finding the shortest paths between all pairs of nodes, and other related problems.

A discussion of stacks and related data structures can be found in reference 3.

3. *The Art of Computer Programming*; vol. I: Fundamental Algorithms, D. E. Knuth, Addison-Wesley, Reading, Mass., 1968.

10.
DIRECTED GRAPHS
AND FLOW NETWORKS

10.1 DIRECTED GRAPHS

In Chapter 7, when we discussed the mathematical notation for an undirected graph, we emphasized that the pair of nodes representing a branch is unordered, and that no particular direction is associated with a branch. We now want to consider *directed graphs,* and each branch will be considered to have a direction associated with it. This feature will enable us to represent the situation where we can send something from node i to node j, say, but not in the reverse direction.

We shall denote the directed branch from node i to node j by (i,j) and the reverse branch by (j,i). We can no longer use these symbols interchangeably, since we now distinguish between two branches that connect the same two nodes but that have different orientations. If the branch (i,j) is in a graph, we shall say that node i *precedes j,* and that node j *succeeds i.* Thus, if branches (i,j) and (j,i) are both present in a graph, node i both precedes and succeeds node j.

The mathematical notation for a directed graph is unchanged from that for an undirected graph: it will consist of a pair of sets

$$G = \{N, B\} \qquad (10.1.1)$$

where N is the set of nodes, and B is the set of directed branches.

As an example of a directed graph, let us return to the airline map considered in Chapter 9. It may happen that airline service provides a flight in only one direction between a given pair of cities. In such a case we can indicate this fact by associating a direction with the branch representing the flight service between those two cities, pointing in the direction the flight takes. This can be indicated on the picture of the graph by an arrow on the branch. If there is two-way service between two cities, we can indicate this by putting two branches between the cities, one oriented in each direction. For the airline schedule of Fig. 9.1.1, we might have the graph shown below in Fig. 10.1.1.

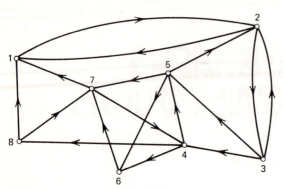

FIG. 10.1.1 A directed graph representing an airline schedule.

As a second example, suppose that the tournament represented by the graph in Fig. 7.3.1 has already been played. We can represent the fact that player *i* beat player *j* by putting a directed branch from node *i* to node *j*, as shown below in Fig. 10.1.2.

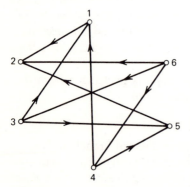

FIG. 10.1.2 A directed graph representing a tournament.

For a third example of a directed graph, consider a set of integers, and connect *i* to *j* if and only if *i* divides *j* evenly. For the integers 1 to 6, the directed graph in Fig. 10.1.3 results (since i divides i, each node has a self-loop):

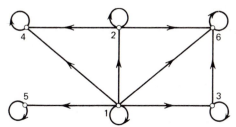

FIG. 10.1.3   Directed graph representing the relation "i divides j."

The gas pipeline collection network shown in Fig. 7.2.3 can be converted into a directed graph in a natural way, simply by orienting each branch in the direction of flow from well to collection point. Similarly, the graph in Fig. 7.2.2 representing the calculations in the FFT algorithm is naturally converted to a directed graph by drawing arrows on each branch pointing to the new quantity being calculated.

It is always possible to convert an undirected graph into a *completely equivalent* directed graph, simply by replacing each undirected branch by two directed branches between the same two nodes, one in each direction. Of course, the reverse is not true; it is not always possible to replace a directed graph by an undirected one. In this sense directed graphs include undirected graphs as special cases.

Example: Signal Flow Graphs

The digital filters discussed in Part I provide us with a concrete example of directed graphs. Consider first the moving average filter defined by

$$Y(k) = A(1)X(k) + A(2)X(k-1) + A(3)X(k-2)$$

The calculations which must be performed to calculate $Y(k)$ directly from this formula are: (1) multiply the constant $A(1)$ by the signal value $X(k)$; (2) multiply $A(2)$ by $X(k-1)$; (3) multiply $A(3)$ by $X(k-2)$; (4) add the results of the three previous steps. This can be represented by the directed graph shown in Fig. 10.1.4.

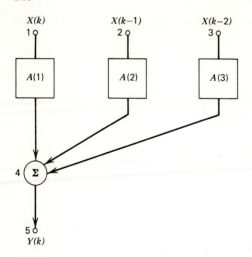

FIG. 10.1.4   First step in the construction of a directed graph to represent a moving average digital filter.

In this graph the signal value $X(k)$ is thought of as residing at node 1. This node can be considered to be a computer register where $X(k)$ is stored, and is called a "register node". Similarly for nodes 2, 3, and 5; which are register nodes for $X(k-1)$, $X(k-2)$, and $Y(k)$, respectively. Node 4 is called a "summing node" and represents the *operation* of summing the values corresponding to its input branches, and passing this sum on to the register node for $Y(k)$, to which its output branch is connected.

Now consider what happens when the time variable k increases by 1. We should then replace the value at node 1, the $X(k)$ register node, by the next value of the signal X; and move the old value of X from node 1 to node 2. This can be represented by inserting a directed branch, called a "delay branch," from node 1 to node 2, with a box containing the delay operation z^{-1}. Figure 10.1.5 shows this delay branch, as well as the delay branch from node 2 to 3.

We have also indicated in this illustration a branch directed in to node 1, representing the source of new values for $X(k)$ (the input); and a branch directed out from node 5, representing the destination of new values of $Y(k)$ (the output).

This graph represents a physical arrangement that could be constructed to implement any moving average digital filter. At an instant when k changes (the time of a "clock" pulse), the contents of register nodes are updated whenever their input branches are delay branches. Then the multiplications by coefficients are performed, and the results added and transferred to the output register node $Y(k)$.

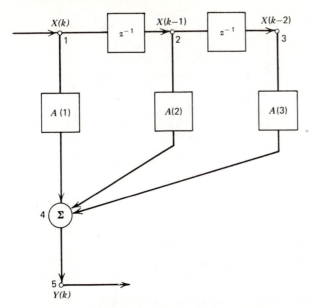

FIG. 10.1.5   Complete directed graph representing a moving average digital filter.

In constructing the signal flow graph above, we have used two kinds of branches: those representing multiplication by constant coefficients, and those representing delay operations; and two kinds of nodes: register nodes and summing nodes. (The branch leading from the summing node 4 to register node 5 in our example has no indication as to what kind it is; by convention, we take this to mean multiplication by unity.)

To implement a recursive digital filter, we need to save some previous output values, $Y(k-1)$, $Y(k-2)$, and so on, in register nodes. Figure 10.1.6 shows the signal flow graph above with two additional register nodes, one for $Y(k-1)$ and one for $Y(k-2)$. Now to represent the recursive digital filter defined by the equation

$$Y(k) = A(1)X(k) + A(2)X(k-1) + A(3)X(k-2)$$
$$+ B(1)Y(k-1) + B(2)Y(k-2)$$

we simply add multiplication branches corresponding to the coefficients $B(1)$ and $B(2)$, as shown in Fig. 10.1.7.

This last signal flow graph can be thought of as being an implementation of the transfer function $[A(1) + A(2)z^{-1} + A(3)z^{-2}]$ followed by an implementation of the transfer function $1/[1 - B(1)z^{-1} - B(2)z^{-2}]$. That is, it implements the numerator first and then the denominator. Figure 10.1.8, on the other hand, shows a signal flow graph that implements the

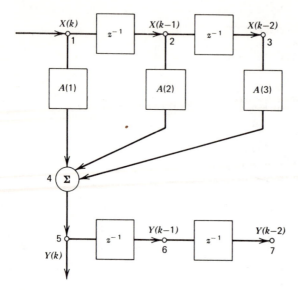

FIG. 10.1.6   The next step in representing a recursive digital filter.

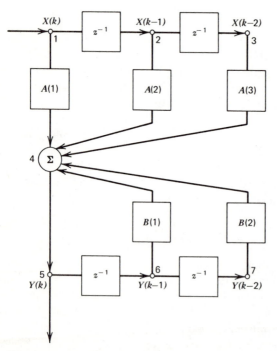

FIG. 10.1.7   Complete directed graph representing a recursive digital filter. The top part of the graph represents the numerator of the transfer function, while the bottom part represents the denominator.

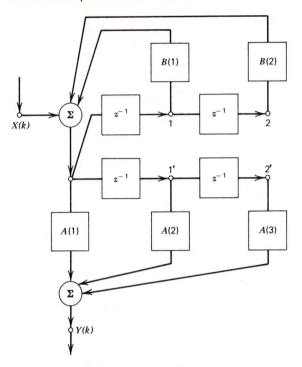

FIG. 10.1.8 Directed graph representing the same recursive digital filter as in the previous figure, but with the denominator at the top and the numerator at the bottom.

same overall transfer function, but in reverse order: denominator followed by numerator.

A simple observation leads to some simplification in this last signal flow graph. Notice that the register nodes marked 1 and 1' always have the same value in them: a delayed version of the output of the same summing node. Similarly for the register nodes marked 2 and 2'. Hence, register nodes 1' and 2' can be eliminated, resulting in the signal flow graph of Fig. 10.1.9.

Exercise 10.1.1

Suppose the graphs given as examples in this section are represented by the notation $\{N,B\}$. Write the set B for each one.

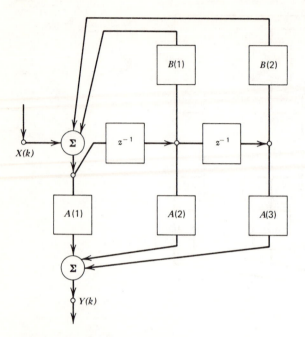

FIG. 10.1.9 Simplified directed graph representing the same recursive digital filter as in the two previous figures.

Exercise 10.1.2

In the tournament represented by the graph in Fig. 10.1.2 who would you say won? How would you settle this question for an arbitrary tournament graph?

Exercise 10.1.3

Generalize Fig. 10.1.9 to show a signal flow graph representing the digital filter with transfer function

$$H(z) = \frac{A(1) + \ldots + A(M) z^{-(M-1)}}{1 - B(1) z^{-1} - \ldots - B(L) z^{-L}}$$

Exercise 10.1.4

What is the transfer function $Y^*(z)/X^*(z)$ represented by the following graph, assuming that the box marked $H(z)$ contains the signal flow graph for the digital filter in the previous exercise:

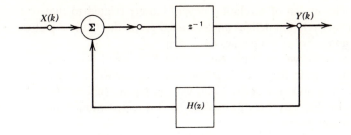

Exercise 10.1.5

What are the transfer functions $Y^*_1(z)/X^*(z)$ and $Y^*_2(z)/X^*(z)$ for the signal flow graph shown below? This implementation of a two-pole one-zero digital filter is called the *coupled form* by Gold and Rader (see the suggestions for further reading at the end of Chapter 6).

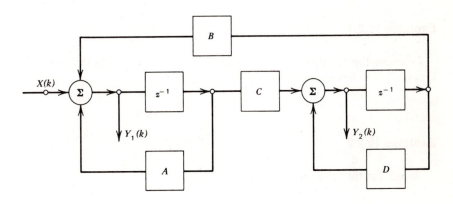

*Exercise 10.1.6

Not all graphs constructed in arbitrary ways using the two types of nodes and branches we have discussed represent well-defined sequences of operations. Give an example of a graph that does not make sense, and discuss the problem of deciding whether a graph does or does not represent a well-defined sequence of operations.

10.2 RELATIONS AND GRAPHS

Often we speak of a relation as holding between certain elements of a set of objects. As an example, consider the set of people and the relation "is father to." The notion of a relation is very closely related to that of a graph. In fact, every relation defines an associated graph that gives a picture of the relation, and every graph defines a relation.

To be more general, we shall use the symbol R to stand for some relation, and we shall write that aRb if a has this relation to b. If we now draw a graph with a node for each object, we can put in the branch (a,b) whenever aRb. In this way we can express by a directed branch the fact that a certain relation holds.

A relation is called *symmetric* if aRb implies that bRa. This notion corresponds exactly to that of an undirected graph. A relation is called *reflexive* if for every element a in our set of objects, aRa. This means that the corresponding graph has a self-loop at every node. Finally a relation is called *transitive* if aRb and bRc together imply that aRc. An example of a transitive relation is provided by the relation between integers "a divides b." Thus the graph of Fig. 10.1.3 is transitive.

Exercise 10.2.1

Consider the relation between people: "is a parent of." Is this relation symmetric? Reflexive? Transitive?

Exercise 10.2.2

Repeat for the relation "is a sibling of."

Exercise 10.2.3

Suppose a relation is symmetric, reflexive, and transitive. What does this imply about its graph?

Exercise 10.2.4

Give an example of a relation that is transitive, other than the ones discussed above. Draw an example of a graph associated with this relation.

10.3 THE STORAGE OF DIRECTED GRAPHS IN A COMPUTER

The methods for graph storage discussed earlier apply equally well to directed graphs, without much modification. Consider first the branch-list method. This carries over without any change at all: we need merely remember that the branch (J,K) is represented by entries $B1(I) = J$ and $B2(I) = K$ on the branch list, and that the order of J and K is now important.

The second method of graph storage, the adjacency-list method, also carries over easily. Instead of the array NEAR(I,J) that contains two entries for every undirected branch, we shall use the array AFTER(I,J), which will contain in its (I,J)th position the Jth node that succeeds I. Thus every directed branch will be represented by exactly one entry in AFTER(I,J). NUMBER(I) will now contain the number of nodes that succeed I. Thus, the directed airline graph shown in Fig. 10.1.1 will be stored as follows:

I	NUMBER(I)	AFTER(I,J), J= 1	2	3	4	5	6	7	8
1	1	2	—	—	—	—	—	—	—
2	2	1	3	—	—	—	—	—	—
3	3	2	5	4	—	—	—	—	—
4	3	5	8	6	—	—	—	—	—
5	3	2	7	6	—	—	—	—	—
6	1	7	—	—	—	—	—	—	—
7	2	4	1	—	—	—	—	—	—
8	2	1	7	—	—	—	—	—	—

$$(10.3.1)$$

The third method of graph storage, the adjacency-matrix method, also carries over easily to directed graphs. We store in ADJ(I,J) a "1" or a "0" according as the directed branch (I,J) is or is not in our graph. The only difference is that now the ADJ(I,J) matrix need not be symmetric. Shown below is the ADJ(I,J) matrix for the airline graph:

I	ADJ(I,J), J=	1	2	3	4	5	6	7	8
1		0	1	0	0	0	0	0	0
2		1	0	1	0	0	0	0	0
3		0	1	0	1	1	0	0	0
4		0	0	0	0	1	1	0	1
5		0	1	0	0	0	1	1	0
6		0	0	0	0	0	0	1	0
7		1	0	0	1	0	0	0	0
8		1	0	0	0	0	0	1	0

The shortest path algorithm described in Chapter 9 works without change for directed graphs. At each stage the distances from labeled nodes to unlabeled nodes that are successors of the labeled nodes should be examined. Thus, if the AFTER array is read into the computer, the program will find a directed path from the origin to the destination node; that is, a path will be found that traverses branches only in the correct direction.

Exercise 10.3.1

Describe another method of adjacency-list storage for directed graphs, based on preceding rather than succeeding nodes.

Exercise 10.3.2

Describe a method for storing in a computer all the information associated with a signal flow graph. (See the example in Section 10.1.)

Exercise 10.3.3

Flow charts for computer programs can be represented by directed graphs. Describe the details of their representation, and give a method for storing in a computer all the information required to describe a flow chart.

10.4 FLOW IN A GRAPH: KIRCHHOFF'S CURRENT LAW

In many physical networks there is a flow of some commodity, such as electricity, water, or information, between nodes. We can represent this by thinking of each branch of a graph as representing a kind of pipe or conduit through which a commodity can flow. The flow in a graph can be stored in a two-dimensional array $F(I,J)$, where we store in $F(I,J)$ the flow from node I to node J. A flow from node I to J can exist, of course, only if the branch (I,J) actually exists in our graph. We shall assume that material cannot flow in the wrong direction in a directed branch. Hence, $F(I,J)$ will be either a positive number or zero. Graphs that support the flow of a commodity are called *flow networks*, and from now on we shall use the term *network* interchangeably with *graph*.

Flow in a network has the fundamental property that material is conserved. Thus, whatever enters a pipe at one end must come out the other end. In addition, there can be no accumulation or deficit of the commodity at a node. We shall now express this fact mathematically, in a statement called *Kirchhoff's current law (KCL)*. Consider a node I that may be the successor of some nodes and the predecessor of some nodes, as shown in Fig. 10.4.1.

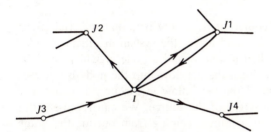

FIG. 10.4.1   A typical node I in a flow network.

The net flow into node I can be obtained by adding the flow in all branches that enter I, and then subtracting the flow in all branches that leave I. Hence the net flow into node I is

$$\text{Net Inflow} = \sum_{\text{branches}(J,I)} F(J,I) - \sum_{\text{branches}(I,J)} F(I,J)$$

If a branch (J,I) is not present in our network, the flow stored in the entry $F(J,I)$ will be zero. Hence, in terms of our F array, we can sum over all J in the summations above, and automatically get all the flow corresponding to actual branches. Thus, we can write the net flow into node I as

$$\text{Net Inflow} = \sum_{\text{all } J} F(J,I) - \sum_{\text{all } J} F(I,J)$$

This second expression is easier to evaluate on a computer, since we do not have to find the predecessors and successors of I. On the other hand, it is a bit less efficient to add a lot of zeros to a sum, when we could avoid these steps. For simplicity, let us use the second formula to compute the net flow into node I:

```
INTEGER F(50,50),INFLO
INFLO=0
DO 1 J=1,N
1 INFLO=INFLO+F(J,I)−F(I,J)
```

Our material conservation law states that the net flow into node I must in fact be zero. Hence

$$\text{Net Inflow} = \sum_{\text{all } J} F(J,I) - \sum_{\text{all } J} F(I,J) = 0 \quad (\text{KCL})$$

and this is called KCL. We could use the computer code above to check KCL in a flow network.

10.5 SOURCES AND SINKS

In a physical situation, of course, it is usually true that material enters the system at one or more nodes, and leaves the system at others. A node at which material enters is called a *source,* and a node at which material leaves, a *sink.* At sources and sinks, the net flow into the node is not zero and KCL does not hold. Instead the net inflow at a source is equal to a negative number, called the *strength* of the source. We can represent this by adding a fictitious branch at a source, coming from outside the graph (Fig. 10.5.1) .

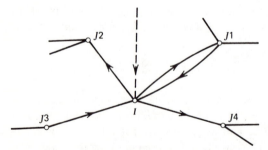

FIG. 10.5.1   Adding a fictitious branch at a source.

We could think of this as a pump injecting material into our network at node I. The computer code above would calculate a net inflow that is a negative number, since we would fail to take the injecting branch into account. Another way of looking at this is to calculate the net outflow

$$\text{Net Outflow} = \sum_{\text{all } J} F(I,J) - \sum_{\text{all } J} F(J,I)$$

which is the negative of the net inflow, and is a positive quantity for a source.

Similarly, the net inflow at a sink is a positive number, which may be called the *strength* of the sink. Again, we may represent a sink by adding a fictitious branch, as shown in Fig. 10.5.2.

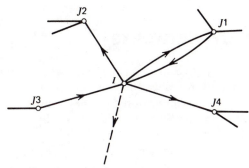

FIG. 10.5.2   Adding a fictitious branch at a sink.

Of course all the material that is pumped into a network at sources must be removed at sinks. Thus the sum of the strengths of all the sources and sinks must equal zero. Consider a network with exactly one source and one sink. Then we can think of the material arriving at the sink as being diverted to the source through a branch outside the graph (see Fig. 10.5.3).

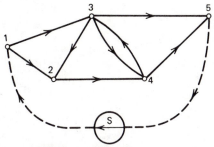

FIG. 10.5.3   Representation of an external pump that collects material at a sink and supplies it at a source of a flow network.

The circle with an *S* inside can be thought of as a pump that collects material at node 5 and pushes it back into node 1. In such a situation, we say that a *flow of value S exists* from node 1 to node 5.

If the fictitious branch from node 5 to node 1 is counted as a real branch in the graph, then KCL holds at all nodes.

Exercise 10.5.1

Give an example of a commodity and a corresponding network, for which KCL does not hold.

10.6 CAPACITATED FLOW NETWORKS

In some flow networks there may be a limit on the amount of a commodity that can be sent through a given branch. Such networks are called *capacitated*. The capacity of each branch can be stored in a two-dimensional array *C* by putting in entry $C(I,J)$ the capacity of branch (I,J). By setting an element of *C* to zero, one effectively removes this branch from the graph. Thus, we can store all the information required to describe a capacitated flow network in the capacity matrix *C*. This does not count the information regarding the flow itself, which is stored in another matrix *F*.

Exercise 10.6.1

Consider the capacitated flow network below, with the capacities shown beside the branches:

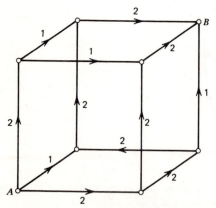

FIG. 10.6.1   Flow graph for Exercise 10.6.1.

What is the most flow that can exist from the node labeled *A* to the node labeled *B*? Write the *C* matrix for the graph in Fig. 10.6.1 and the *F* matrix corresponding to your solution.

Exercise 10.6.2

Repeat for the graph of Fig. 10.6.2.

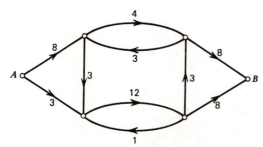

FIG. 10.6.2   Flow graph for Exercise 10.6.2.

Exercise 10.6.3

Repeat for the graph of Fig. 10.6.3.

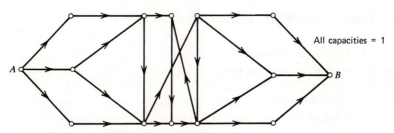

FIG. 10.6.3   Flow graph for Exercise 10.6.3.

Exercise 10.6.4

Consider the problem of finding a path in a capacitated network from a given node to another given node, which has maximum capacity in the sense that the smallest capacity branch on the path is as large as possible. Describe a labeling algorithm for solving this problem.

Exercise 10.6.5

All the branches in Fig. 10.6.4 are undirected, and the numbers indicate the capacity in either direction. Find the path from x to y with the largest capacity in the sense of Exercise 10.6.4.

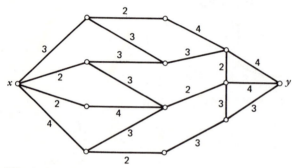

FIG. 10.6.4   Flow graph for Exercise 10.6.5.

10.7   THE MAXIMUM FLOW PROBLEM

A problem that arises naturally in the study of capacitated flow networks is that of assigning flows to the branches of the network so as to maximize the net flow from a given source node to a given sink node. This problem is called the *maximum flow* problem. Its solution for a network of some complexity is not usually obvious; yet, we shall show that the problem can be solved readily by repeated application of a simple labeling algorithm.

We shall use the network in Fig. 10.7.1 to illustrate what we say as we go along. Node 4 will be designated as the source, and node 5 as the sink. The capacity of each branch is shown beside the branch in a circle. When we assign a flow to a branch, we shall indicate this by putting an uncircled number beside the branch. Thus, we can send a flow of 1 unit from node 4 to 1 to 3 to 5 as shown in Fig. 10.7.2.

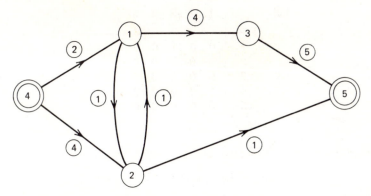

FIG. 10.7.1 Flow graph used to illustrate the maximum flow problem. Circled numbers represent branch capacities.

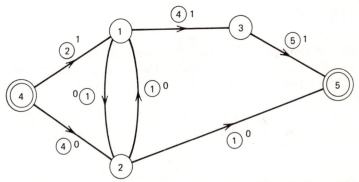

FIG. 10.7.2 The initial flow pattern in the example; uncircled numbers represent flow values.

The capacity matrix of this network will be stored in the computer in the two-dimensional array shown below:

C(I,J)	1	2	3	4	5
1	0	1	4	0	0
2	1	0	0	0	1
3	0	0	0	0	5
4	2	4	0	0	0
5	0	0	0	0	0

As mentioned in Section 10.6 this describes the network completely. The flow shown in Fig. 10.7.2 will be stored in the flow array as shown below:

F(I,J)	1	2	3	4	5
1	0	0	1	0	0
2	0	0	0	0	0
3	0	0	0	0	1
4	1	0	0	0	0
5	0	0	0	0	0

Since the flow in a branch cannot exceed its capacity, we must always have

$$0 \leq F(I,J) \leq C(I,J)$$

for every I and J.

10.8 CUTS

A fundamental concept in the study of networks is that of a *cut*. An $I1$–$I2$ cut is defined as a partition of the nodes of the network into two sets, V and $\overline{V}$, such that $I1$ is in V and $I2$ is in $\overline{V}$. We draw a cut by dividing the nodes into two groups, V and $\overline{V}$, and showing the branches that go between V and $\overline{V}$. For example, a 4-5 cut for our illustrative network can be drawn by choosing nodes 4, 2, and 3 for V, as shown in Fig. 10.8.1.

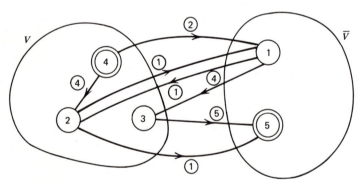

FIG. 10.8.1   A 4-5 cut for the example.

Notice that all the flow going from a source in V to a sink in $\overline{V}$ must pass through the branches that bridge these two sets of nodes. This is limited by the capacity of all the branches directed from V to $\overline{V}$, since we can do no better than to saturate those branches directed from V to $\overline{V}$.

The sum of the capacities of all the branches directed from V to $\overline{V}$ is called the *capacity of the cut* and is denoted by $C(V,\overline{V})$. Thus

Definition

$$C(V,\overline{V}) = \sum_{\substack{I\varepsilon V \\ J\varepsilon\overline{V}}} C(I,J)$$

We have shown that the flow from 4 to 5 is bounded by the capacity of any 4-5 cut; that is, in general

$$\text{Flow From } I1 \text{ to } I2 \le C(V,\overline{V})$$

where V and $\overline{V}$ define any *I1-I2* cut.

In the example above, the capacity of the 4-5 cut shown is 9, so that

$$\text{Flow From 4 to 5} \le 9$$

This upper bound is clearly not attainable here, since it is not possible for there to be a flow of 5 from node 3 to 5 while there is no flow at all from node 1 to 3, since then node 3 would be a source.

10.9 AUGMENTATION PATHS

Let us go back to the flow of value 1 shown in Fig. 10.7.2 and consider how we might push some more flow from node 4 to 5. We can send another unit of flow from node 4 to 1, and this can be diverted down to node 2 and over to node 5, resulting in a flow of 2 units as shown in Fig. 10.9.1.

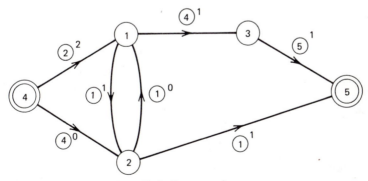

FIG. 10.9.1   A flow of 2 units in the example.

Branch (4,1) is now saturated, but we can send another unit of flow from node 4 to 2. This can be sent up to node 1 and across to nodes 3 and 5 resulting in a flow of 3 units (Fig. 10.9.2).

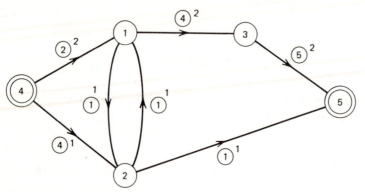

FIG. 10.9.2   A flow of 3 units in the example.

We have now reached a crucial point in our attempt to push flow from 4 to 5. We cannot send any more flow from 4 to 1, since branch (4,1) is saturated. We can try to send another unit down to node 2, however. But the two branches directed out from node 2 are saturated. Our only hope is to cancel a unit of the flow already coming in to node 2. That is, reducing the flow from node 1 to 2 by 1 unit is equivalent to sending 1 unit of flow from 2 to 1, *provided there is already some backward flow to cancel.* This unit of flow can be sent on to nodes 3 and 5, resulting in the flow of value 4 (Fig. 10.9.3).

We can now show that this is the maximum flow possible from node 4 to 5, by considering the 4-5 cut with nodes 4 and 2 in V (Fig. 10.9.4).

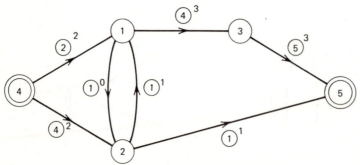

FIG. 10.9.3   A flow of 4 units in the example.

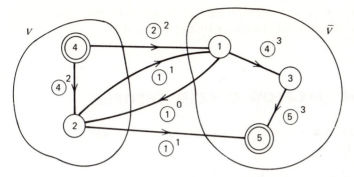

FIG. 10.9.4 A 4-5 cut for the example, showing that the flow pattern is optimal.

The capacity of this cut is 4, hence

$$\text{Flow From 4 to } 5 \leq 4$$

But we have achieved a flow of 4, which means that this is the maximum 4-5 flow possible. Notice that for the cut shown in Fig. 10.9.4 every forward branch (from V to $\overline{V}$) is saturated, and every reverse branch (from $\overline{V}$ to V) is empty.

At each stage of building up the flow, we sent flow along a path from the source node to the sink node. This path had the property that each branch used had either residual capacity in the forward direction or, as in the last path, nonzero flow in the reverse direction. For example, the fourth unit of flow that was added to the network in Fig. 10.9.2 to get the flow shown in Fig. 10.9.3 was sent along the path shown in Fig. 10.9.5.

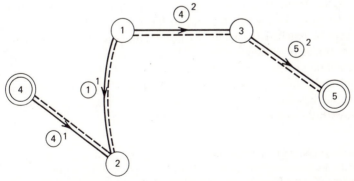

FIG. 10.9.5 The augmentation path used to achieve the maximum flow in the example.

Such a path is called an *augmentation path*, since the flow can be augmented from source to sink if such a path can be found.

10.10 THE MAX-FLOW MIN-CUT THEOREM

We now are in a position to demonstrate a rather remarkable fact: that in the inequality

$$\text{Flow From } I1 \text{ to } I2 \leq C(V,\overline{V})$$

equality can always be achieved for some $I1$-$I2$ cut. Thus,

$$\text{Maximum Flow From } I1 \text{ to } I2 = \underset{\substack{I1\varepsilon \underline{V} \\ I2\varepsilon \overline{V}}}{\text{Minimum}} C(V,\overline{V})$$

This is known as the *max-flow min-cut* theorem, and is a fundamental property of flow networks.

To show this, we should first notice that the problem of finding an augmentation path is precisely the problem of finding any path from node 4 to 5 in a network that has branches of finite length only where the original network has either a forward branch that is unsaturated, or a reverse branch that has nonzero flow. As we discussed in Chapter 9 this problem can be solved by applying the labeling algorithm, starting with the source node, and at each step labeling all unlabeled nodes that can be reached from labeled nodes. A labeling can legitimately take place from node I to node J if (1) branch (I,J) is unsaturated, or (2) branch (J,I) has nonzero flow. Each time an augmentation path is found, the flow can be augmented along it, backtracking from the sink to the source.

Eventually, the point is reached where no augmentation is found by the labeling process.* In the last attempt to label, no unlabeled nodes are found that can be reached from labeled nodes through forward branches that are unsaturated or through reverse branches that have nonzero flow. Hence if we define at this point a cut with the labeled nodes in set V, and the unlabeled nodes in set $\overline{V}$, all the branches from V to $\overline{V}$ are saturated, and all the branches from $\overline{V}$ to V are empty. This situation is shown in Fig. 10.10.1, with the labeled nodes indicated by asterisks.

---

* The process is guaranteed to terminate in a finite number of steps only if the capacities $C(I,J)$ are integers. See reference 1 in the suggestions for further reading at the end of this chapter for a non-terminating example with irrational capacities, reference 2 for a way around the difficulty, and Exercise 10.12.1.

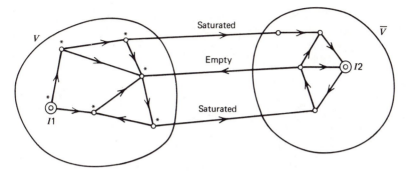

FIG. 10.10.1   The final stage in the labeling process, when no augmentation path can be found.

Thus for this cut,

$$\text{Flow From } I1 \text{ to } I2 = C(V, \overline{V})$$

where V and $\overline{V}$ define an $I1$-$I2$ cut. But the flow from $I1$ to $I2$ is bounded by the capacity of any $I1$-$I2$ cut, so this must, in fact, be the maximum flow possible from $I1$ to $I2$. This proves the *max-flow min-cut* theorem.

This result can be interpreted intuitively as saying that the largest flow is determined by the narrowest bottleneck, where we interpret the branches across a cut as a bottleneck through which all flow must pass. The labeling algorithm for building up the maximum flow was first suggested by Ford and Fulkerson (see the suggestions for further reading).

10.11 PROGRAMMING THE FORD AND FULKERSON LABELING ALGORITHM

We shall now fill in the details of the labeling algorithm described above. At each application of the labeling algorithm, some flow has been assigned to the branches, and we are seeking an augmentation path from $I1$ to $I2$. We shall use a two-part label. The first label, $L1(J)$, plays the same role as in the shortest path algorithm; that is, $L1(J)$ will be 0 if J is unlabeled, and will contain the number of the node from which we arrived at node J if J is labeled. The second label, $L2(J)$, will contain the amount by which the flow can be augmented. This is obtained from the previous $L2$ as follows: suppose the previous node is I. Then there are two possible ways we could have reached J. First, we could have reached J by labeling forward along an unsaturated branch. In this case the increment of flow is the smaller of $L2(I)$, the second label of the preceding node, and $C(I,J) - F(I,J)$, the residual capacity in the branch from I to J. This situation is illustrated in Fig. 10.11.1.

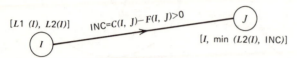

FIG. 10.11.1   Labeling forward along an unsaturated branch.

The second way we could have reached J is by labeling backward along a branch with nonzero flow. In this case the increment of flow is the smaller of $L2(I)$ and $F(J,I)$, the nonzero backward flow. This situation is shown in Fig. 10.11.2.

FIG. 10.11.2   Labeling backward along a branch with nonzero flow.

Notice that in the second case, the $L1$ label of J is set equal to $-I$. This minus sign indicates that this is a backward branch along the augmentation path. This information will be used in augmenting the flow in the back-tracking procedure.

As pointed out in the discussion of the any-path algorithm in Section 9.5, once we label from a node we need no longer try to label from it again. We can use this information by defining a one-dimensional STATE array that tells us whether a node has been labeled to or labeled from. If a node has been labeled from, we say that it is *scanned*. Thus, in each application of the labeling algorithm, a node is in one of three states: unlabeled, unscanned (STATE(I) = 1); labeled, unscanned (STATE(I) = 2); or labeled and scanned (STATE(I) = 3). We can apply a label only to a node whose STATE is 1. After labeling a node, its STATE is set equal to 2. After labeling from a node, its STATE is set equal to 3.

When we label the sink node $I2$, we backtrack, augmenting the flow as we go. The amount of the augmentation is determined by $L2(I2)$. By maintaining two labels, we have simultaneously determined an augmentation path and an increment $L2(I2)$ by which we can augment the flow along the path. Once the flow is augmented, we set the labels to 0, the STATE vector to 1, and start the labeling process over again. Eventually we reach the situation depicted in Fig. 10.10.1 at which point we have found the maximum flow from $I1$ to $I2$.

The complete algorithm can be conveniently divided into two parts: ROUTINE A, the labeling procedure for finding an augmentation path; and ROUTINE B, the augmentation process. An outline of the complete algorithm is given below.

INITIALIZATION: Set all flows to zero.

ROUTINE A: 1. Set all states to 1 (unlabeled, unscanned).
2. Label source $I1$ with $[-1, 100000]$, and set state of $I1$ to 2 (labeled, unscanned).
3. Let I be any labeled unscanned node, initially $I1$.
4. Scan node I, setting states of newly labeled nodes to 2, and then setting state of I to 3 (labeled, scanned).
5. If node $I2$ has become labeled, go to ROUTINE B.
6. Search for a labeled unscanned node I. If one is found, go to 3. If none is found, calculate total flow and return (finished).

ROUTINE B: Backtrack from $I2$ to $I1$, changing flow along the augmentation path. Then go to ROUTINE A.

10.12 A COMPUTER PROGRAM

Figure 10.12.1 shows a complete computer program for performing the flow maximization. The program is written as a simple main program for test purposes, and a subroutine called MINCUT(I1,I2,I3). The argument I1 designates the source node, I2 the sink node, and the value of the maximum flow is returned as I3. The information describing the network is contained in a COMMON block called FORD. This information is set up by the main program that calls MINCUT. Thus, COMMON contains initially the capacity matrix C, and the number of nodes M. The other arrays in FORD are used as temporary storage for the label arrays, the state array, and the flow matrix F, initially zero.

Statement 19 is the beginning of the labeling procedure, called ROUTINE A. At statement 21, I is a labeled unscanned node, initially set to I1, the source. The DO-loop ending at statement 22 labels all the nodes J possible through unsaturated forward branches. The next DO-loop, ending at statement 23, labels all the nodes J possible through reverse branches with non-zero flow. I is then marked scanned, by setting STATE(I) = 3, and the STATE of I2 examined. If I2 is labeled, we transfer to statement 25, called ROUTINE B, which is the augmentation procedure. If I2 is unlabeled, we search for a labeled unscanned node I. If we find one, we transfer back to statement 21, where we try to label from I. If not, we have found the maximum flow. This is calculated as I3, and we transfer back to the calling program.

ROUTINE B, the augmentation procedure, proceeds by examining L1 (NEXT) at each step. If it is negative, we know that the next node to backtrack to is LAST = −L1(NEXT). Here we *decrease* F(NEXT,LAST) by

```
C......TEST PROGRAM FOR MINCUT
       IMPLICIT INTEGER (A-Z)
       COMMON/FORD/F(30,30),C(30,30),L1(30),L2(30),STATE(30),M
       M=5
       DO 10 I=1,M
       DO 10 J=1,M
    10 C(I,J)=0
       C(1,2)=1
       C(1,3)=2
       C(1,4)=3
       C(2,3)=3
       C(3,2)=2
       C(3,4)=2
       C(4,3)=3
       C(2,5)=3
       C(3,5)=1
       C(4,5)=2
       I1=1
       I2=5
       CALL MINCUT(I1,I2,I3)
       WRITE(6,11)
    11 FORMAT('1CAPACITY MATRIX:')
       DO 12 I=1,M
    12 WRITE(6,15) (C(I,J),J=1,M)
       WRITE(6,13)
    13 FORMAT(' FLOW MATRIX:')
       DO 14 I=1,M
    14 WRITE(6,15) (F(I,J),J=1,M)
    15 FORMAT(' ',30I3)
       WRITE(6,16)I1,I2,I3
    16 FORMAT(' FLOW FROM ',I3,' TO ',I3,' IS ',I3)
       STOP
       END
```

FIG. 10.12.1 A FORTRAN program for finding the maximum flow between two nodes of a flow network. The main program sets up a test problem and calls SUBROUTINE MINCUT, which executes the Ford and Fulkerson labeling algorithm.

```
      SUBROUTINE MINCUT(I1,I2,I3)
      IMPLICIT INTEGER (A-Z)
      COMMON/FORD/F(30,30),C(30,30),L1(30),L2(30),STATE(30),M
C......THIS PROGRAM FINDS THE MAXIMUM FLOW FROM I1 TO I2 USING THE
C......FORD AND FULKERSON LABELING ALGORITHM
C......STATE= 1 IF UNLABELED,UNSCANNED...2 IF LABELED,UNSCANNED...
C......3 IF BOTH LABELED AND SCANNED
C......CLEAR THE FLOW MATRIX F
      DO 18 I=1,M
      DO 18 J=1,M
   18 F(I,J)=0
C......ROUTINE A: FIND AUGMENTATION PATH
C......INITIALIZE STATES TO 1
   19 DO 20 I=1,M
   20 STATE(I)=1
C......LABEL I1
      L1(I1)=-1
      L2(I1)=100000
      STATE(I1)=2
      I=I1
C......I IS A LABELED, UNSCANNED NODE; SCAN FORWARD
   21 DO 22 J=1,M
      INC=C(I,J)-F(I,J)
      IF(STATE(J).NE.1.OR.INC.LE.0)GOTO22
      L1(J)=I
      L2(J)=MINO(L2(I),INC)
      STATE(J)=2
   22 CONTINUE
C......SCAN BACKWARD
      DO 23 J=1,M
      INC=F(J,I)
      IF(STATE(J).NE.1.OR.INC.LE.0)GOTO23
      L1(J)=-I
      L2(J)=MINO(L2(I),INC)
      STATE(J)=2
   23 CONTINUE
      STATE(I)=3
C......BREAKTHROUGH?
      IF(STATE(I2).EQ.2)GOTO25
C......LOOK FOR A NODE IN STATE 2
      DO 24 I=1,M
      IF(STATE(I).EQ.2)GOTO21
   24 CONTINUE
C......NONE FOUND, FINISHED, CALCULATE FLOW AND RETURN
      I3=0
      DO 91 JJ=1,M
   91 I3=I3+F(JJ,I2)-F(I2,JJ)
      RETURN
C......ROUTINE B: AUGMENT FLOW
   25 NEXT=I2
      INC=L2(I2)
   26 LAST=L1(NEXT)
      IF(LAST.LT.0)GOTO27
      F(LAST,NEXT)=F(LAST,NEXT)+INC
      GOTO28
   27 LAST=-LAST
      F(NEXT,LAST)=F(NEXT,LAST)-INC
   28 NEXT=LAST
      IF(NEXT.EQ.I1)GOTO19
      GOTO26
      END
```

the flow increment. If L1(NEXT) is positive, we know that the next node to backtrack to is LAST = L1(NEXT). We then *increase* F(LAST,NEXT) by the flow increment. When we finally backtrack to I1, we transfer back to the beginning of the labeling procedure, statement 19.

Figure 10.12.2 shows the output when the program is run on a test problem, and Fig. 10.12.3 shows a picture of the test flow graph. The capacities of the branches are shown as circled numbers, and the actual flow values are shown as uncircled numbers next to the corresponding capacity values.

```
CAPACITY MATRIX:
  0   1   2   3   0
  0   0   3   0   3
  0   2   0   2   1
  0   0   3   0   2
  0   0   0   0   0
FLOW MATRIX:
  0   1   2   3   0
  0   0   0   0   3
  0   2   0   0   1
  0   0   1   0   2
  0   0   0   0   0
FLOW FROM    1 TO    5 IS    6
```

FIG. 10.12.2 Output of the program shown in the previous figure when run on the test problem.

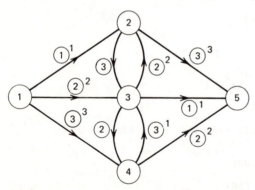

FIG. 10.12.3 Test flow network for the program. The circled numbers represent branch capacities and the uncircled numbers represent the flow values in the solution, a flow of 6 units from node 1 to 5.

Example

We now work through by hand the max-flow problem shown below, where the capacity of every branch is 1. This problem is contrived to illustrate labeling backward and cancelling flow.

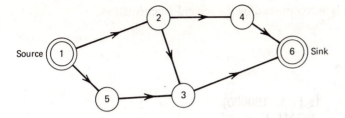

Each search for a flow augmenting path will be called an AUGMENTA-TION. Each time a node is scanned we shall write SCAN I, where I is the node scanned; followed by a list of the newly labeled nodes with their labels. Thus

Augmentation 1

> 1: [−1,100000]
> SCAN 1
> 2: [1,1]
> 5: [1,1]
> SCAN 2
> 3: [2,1]
> 4: [2,1]
> SCAN 5
> SCAN 3
> 6: [3,1]

We now have found the augmentation path back from 6 to 3 to 2 to 1. Augmenting by 1 unit of flow along this path produces the flow pattern shown below, where the augmenting path is made bold, and the flows in branches are indicated.

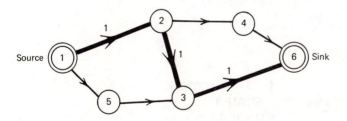

Notice that this path blocks flow along the lower path 1 to 5 to 3 to 6.

The unblocking is accomplished by the second augmentation.

Augmentation 2

1: [−1, 100000]
SCAN 1
5: [1,1]
SCAN 5
3: [5,1]
SCAN 3
2: [−3,1] (a backward label)
SCAN 2
4: [2,1]
SCAN 4
6: [4,1]

The augmentation path is 6 to 4 to 2 to 3 to 5 to 1, with flow being cancelled on the branch (2,3). This augmentation path and the resulting flow pattern is shown below.

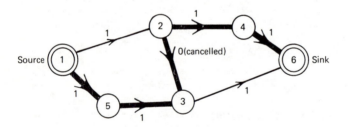

The final augmentation proves, in effect, that this flow of value 2 is maximum.

Augmentation 3

1: [−1, 100000]
SCAN 1
STUCK !

The min-cut consists of the set of all labeled nodes when stuck, that is, the set containing only node 1.

Exercise 10.12.1

Show that if the capacities are all integers, the Ford and Fulkerson labeling algorithm terminates in a finite number of steps. What do you think happens when the capacities are all rational numbers? Irrational numbers?

Exercise 10.12.2

Prove the *integrity theorem*: the maximum flow in a capacitated network with integer capacities is an integer.

Exercise 10.12.3

Solve Exercises 10.6.1, 10.6.2, and 10.6.3 by using the labeling algorithm.

Exercise 10.12.4

We could find the maximum flow by examining all $I1$-$I2$ cuts and choosing the one with the smallest capacity. Show that this is not practical.

Exercise 10.12.5

Find the maximum flow from A to B in the capacitated flow network shown below, using the Ford and Fulkerson labeling algorithm. Find a cut of minimum capacity.

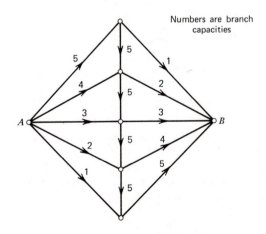

Numbers are branch capacities

Exercise 10.12.6 (computer experiment)

Rewrite subroutine MINCUT so that it keeps a stack of nodes in state 2. Discuss the relative efficiency of this approach and the original method, which is to search for a node in state 2.

Exercise 10.12.7

One criterion of a good test program is that it should cause every possible program statement to be executed. Is this true of the test problem given? If not, determine which statements are not executed and invent a problem that does exercise every statement.

10.13 THE COMPUTER DATING PROBLEM: A MATCHING PROBLEM

We shall now give an example of a problem that can be solved using the maximum flow labeling algorithm. This problem is in the general class called *matching* problems; it involves matching objects in one set with objects in another set. Consider the problem faced by a computer dating service. For each boy applicant, there is a list of girls who are deemed compatible with this boy. This compatibility is determined by questionnaires. In an overly simple example, we might have boys 1, 2, 3, and girls 4, 5, 6; with the following compatibility list:

Boy	Compatible girls
1	4,5
2	5,6
3	4,5

The object is to set up the maximum number of dates, since the dating service charges for each one.

The flow graph in Fig. 10.13.1 represents this situation. Node 7 is a fictitious source, and node 8 a fictitious sink. Every branch has capacity 1. Now a flow from 7 to 8 passes through a boy node, then through a girl node, and then onto the sink. Since each boy has only one incoming branch, and since its capacity is 1, every unit of flow from 7 to 8 must pass through a different boy node and, hence, represents a date. If we can maximize the flow in this network, we then maximize the number of dates.

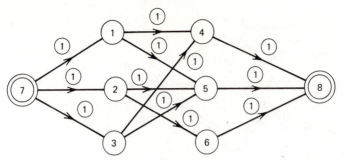

FIG. 10.13.1 Flow network representing a computer dating problem. Circled numbers represent branch capacities.

Applying the labeling algorithm, we can establish 2 units of flow immediately, let us say from 7 to 1 to 4 to 8, and 7 to 2 to 5 to 8 (Fig. 10.13.2).

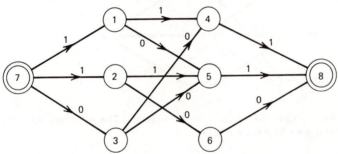

FIG. 10.13.2 A flow pattern of 2 units in the example. The uncircled numbers represent flow values.

This corresponds to boy 1 dating girl 4, and boy 2 dating girl 5. This leads to a frustrating situation for boy 3 and girl 6, since they are the only people left but are incompatible.

The labeling algorithm can rectify this situation as follows: label node 3 from 7 with the two-part label [7,1], the second label being the unit increment of flow. Next label node 4 from 3 with [3,1] and node 5 from 3 with [3,1]. We can now back up branch (2,5) (tentatively break the date between 2 and 5), and label node 2 from 5 with [−5,1]. Next, label node 6 from 2 with [2,1], and finally node 8 from 6 with [6,1]. The labels and the flow augmenting path are shown in Fig. 10.13.3.

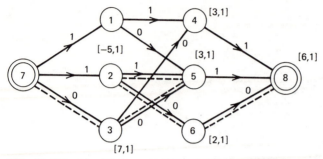

FIG. 10.13.3 The labels and flow augmenting path that lead to a flow pattern of 3 units in the example.

Augmenting the flow according to this augmentation path results in the solution of 3 dates shown in Fig. 10.13.4.

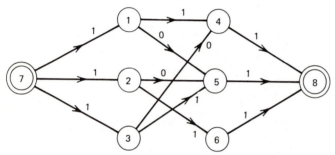

FIG. 10.13.4 Final, optimal, flow pattern in the example. The uncircled numbers represent a flow of 3 units from node 7 to 8.

Exercise 10.13.1

There are 6 men, each of whom can operate 2 of 6 possible machines. The men are numbered 1 to 6, the machines 7 to 12. The machines each man can work are shown below:

Man	Machine	
1	7	8
2	10	11
3	9	10
4	7	11
5	12	10
6	12	9

Find an assignment of men to machines that keeps the most men busy.

Further Reading

The Ford and Fulkerson maximum flow labeling algorithm is discussed in the following books.

1. *Flows in Networks,* R. Ford, Jr., and D. R. Fulkerson, Princeton University Press, Princeton, New Jersey, 1962.

2. *Integer Programming and Network Flows,* T. C. Hu, Addison-Wesley, Reading, Mass., 1969.

3. *Communication, Transmission, and Transportation Networks,* H. Frank and I. T. Frisch, Addison-Wesley, Reading, Mass., 1971.

These books also discuss related flow problems, including the maximization of flows of two and more commodities.

11.
ELECTRICAL NETWORKS

11.1 VOLTAGE, CURRENT, AND RESISTANCE

Electrical networks are flow networks in which the commodity is charge. The flow of charge is called *current*. Electrical networks differ from the capacitated networks discussed before in a fundamental respect: the distribution of flow among the branches is determined uniquely by physical laws, rather than being determined rather freely by our choice. Thus, in an airline network, we are free to send any number of airplanes along any route, provided only that we observe the limits imposed by the capacities. In an electrical network, the current distribution is determined by the properties of the materials used.

Consider, for example, the connection of a battery across a conductor. The battery maintains a potential difference, called a *voltage,* across the conductor, thereby establishing an electric field within the conductor. This electric field produces a force on any free charges (electrons) in the conductor, causing a current to flow. This situation is shown in Fig. 11.1.1, where the battery maintains a voltage of V volts.

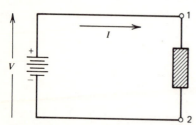

FIG. 11.1.1   A battery connected to a conductor.

The conductor is characterized by the relationship between the applied voltage and the resulting current. A typical voltage-current relationship is illustrated in Fig. 11.1.2.

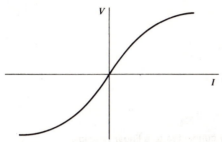

FIG. 11.1.2   Nonlinear relationship between voltage and current.

A negative current corresponds to a charge flow opposite to that shown in Fig. 11.1.1. Thus, the network shown is really an undirected graph, and the arrow associated with I indicates the direction of flow we choose to call positive. Similarly, the arrow associated with V indicates the potential difference we choose to call positive. In other words, if the potential at node 1 is higher than that at node 2, V is a positive number.

An important kind of conductor is one in which the current is proportional to the voltage. Such a conductor is called a *linear resistor,* and the voltage-current curve in this case is a straight line (Fig. 11.1.3).

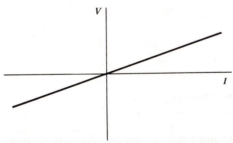

FIG. 11.1.3   Linear relationship between voltage and current.

This relationship occurs in certain materials and is called *Ohm's law.*

In many materials Ohm's law does not hold. An example is an ionized gas in which the free charges are ions as well as electrons. Transistors and vacuum tubes provide other examples of nonlinear voltage-current relationships. In this chapter we shall restrict our attention to linear resistors (called simply *resistors* when there is no danger of ambiguity). A linear resistor will be represented by a zig-zag line (Fig. 11.1.4).

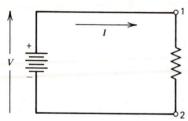

FIG. 11.1.4   Representation of a battery connected to a linear resistor.

The battery is an example of a device that maintains a constant voltage across its terminals. Such a device is called a *voltage source*. Generators and dry cells provide further examples.

We can also postulate the existence of a *current source*: this device supplies a constant current through its terminals, regardless of external conditions. Such a device can be thought of as a current pump, and is represented by the symbol in Fig. 11.1.5.

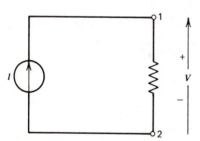

FIG. 11.1.5   Representation of a current source.

The situation is now reversed: we must now determine the voltage from the current.

The ratio of voltage to current in a linear resistor is called *resistance*, and is usually represented by the letter R, possibly with subscripts. Thus, in the networks above, $V = IR$. The reciprocal of R is called *conductance*, and is usually represented by the letter G. The units of resistance are *ohms*, and of conductance, *mhos*.

11.2 THE EQUIVALENCE BETWEEN VOLTAGE AND CURRENT SOURCES

Consider a voltage source connected to a resistor (Fig. 11.2.1).

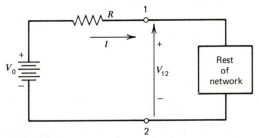

FIG. 11.2.1   A voltage source and resistor in some network.

The nodes 1 and 2 may be connected to some other electrical devices. The voltage across R is $V_0 - V_{12}$, so by Ohm's law, the current I is

$$I = \frac{V_0 - V_{12}}{R} \tag{11.2.1}$$

Now consider a current source connected across a resistor as shown in Fig. 11.2.2.

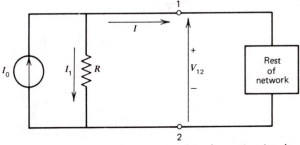

FIG. 11.2.2   A current source and resistor in some network.

The current through R is, by KCL, $I_1 = I_0 - I$, so that

$$V_{12} = (I_0 - I) R \tag{11.2.2}$$

If we solve for I, we have

$$I = \frac{I_0\, R - V_{12}}{R} \tag{11.2.3}$$

Notice that Eqs. 11.2.1 and 11.2.3 both express a relationship between V_{12}, the voltage difference between nodes 1 and 2; and I, the current out of node 1 and into node 2. Notice further that these two expressions are identical, provided that the values of R are the same, and provided that $V_0 = I_0 R$. Thus, *as far as the rest of the network is concerned*, these two situations are identical, and one may be replaced freely by the other. The networks of Figs. 11.2.1 and 11.2.2 are said to be *equivalent networks*. They are equivalent only as regards things outside the nodes 1 and 2.

As we mentioned, certain devices, such as batteries, are represented naturally by voltage sources. These devices almost always have associated with them some unavoidable internal resistance, and we shall assume in the following sections that every voltage source has in series with it some positive resistance. This means that we can always perform the conversion from voltage to current source described above, and we shall therefore restrict our attention to networks that contain only current sources. (The assumption that every voltage source has a series resistor is not necessary, but simplifies this discussion.)

Exercise 11.2.1

Find a network with a current source that is equivalent to the network in Fig. 11.2.3 in regard to everything outside terminals 1 and 2.

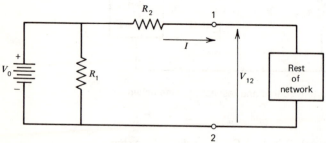

FIG. 11.2.3   Network for Exercise 11.2.1.

Exercise 11.2.2

Repeat for the network in Fig. 11.2.4.

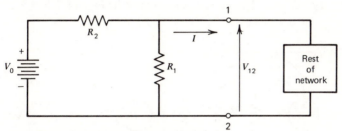

FIG. 11.2.4   Network for Exercise 11.2.2.

Exercise 11.2.3

Show that the networks in Fig. 11.2.5 are equivalent in regard to everything outside terminals 1 and 2.

FIG. 11.2.5   Networks for Exercise 11.2.3.

Exercise 11.2.4

Repeat for the networks in Fig. 11.2.6.

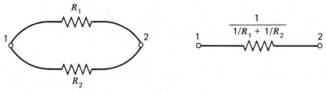

FIG. 11.2.6   Networks for Exercise 11.2.4.

11.3 RESISTIVE NETWORKS AND NODE EQUATIONS

As we have seen, we can represent any network of voltage sources, current sources, and resistors by a network without voltage sources. We shall now discuss some conventions for assigning voltages and node numbers. For this purpose, consider the resistor network in Fig. 11.3.1.

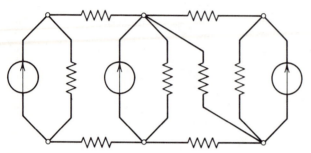

FIG. 11.3.1   Resistor network used as an example.

Since we are interested only in potential difference, we can arbitrarily assign to some node a zero potential. This node will be numbered as node "0," and will be called the *datum* or *ground* node. All voltages will be measured with respect to this node. The other nodes will simply be numbered from 1 to N in some convenient order. The result is shown with the datum node grounded (Fig. 11.3.2).

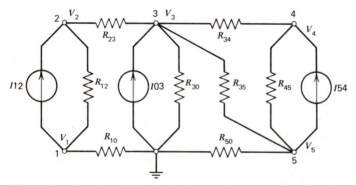

FIG. 11.3.2   Assignment of symbols in the example resistor network.

The variable V_i designates the voltage of node i with respect to the ground, and R_{ij} designates the value of resistance between nodes i and j. If we can find the N voltages $V_1, \ldots, V_N$, we can then find the current in any resistor simply by applying Ohm's law. Thus, the current in R_{12} is simply $(V_2 - V_1)/R_{12}$; with flow being positive in the direction from node 2 to 1. If the original network contained parallel or series resistors, these can be combined into equivalent single resistors, as described in Exercises 11.2.3 and 11.2.4.

Consider now KCL at each node. It states that the net flow out of node 1, for example, is zero. Since each current leaving node 1 is either due to a known current source, or can be expressed in terms of the node voltages of the network, KCL for node 1 yields an equation involving the node voltages, and other quantities, all of which are known:

$$\text{Net flow out of node } 1 = V_1 \left(\frac{1}{R_{10}}\right) + (V_1 - V_2)\left(\frac{1}{R_{12}}\right) + I12 = 0$$

Writing KCL at nodes 1 through N, we arrive at the following set of equations:

$$V_1\left(\frac{1}{R_{10}}\right) + (V_1 - V_2)\left(\frac{1}{R_{12}}\right) + I12 = 0$$

$$(V_2 - V_1)\left(\frac{1}{R_{12}}\right) + (V_2 - V_3)\left(\frac{1}{R_{23}}\right) - I12 = 0$$

$$(V_3 - V_2)\left(\frac{1}{R_{23}}\right) + V_3\left(\frac{1}{R_{30}}\right) + (V_3 - V_5)\left(\frac{1}{R_{35}}\right) +$$
$$(V_3 - V_4)\left(\frac{1}{R_{34}}\right) - I03 = 0$$

$$(V_4 - V_3)\left(\frac{1}{R_{34}}\right) + (V_4 - V_5)\left(\frac{1}{R_{45}}\right) - I54 = 0$$

$$V_5\left(\frac{1}{R_{50}}\right) + (V_5 - V_3)\left(\frac{1}{R_{35}}\right) + (V_5 - V_4)\left(\frac{1}{R_{45}}\right) + I54 = 0$$

We now have N equations in the N unknowns $V_1, \ldots, V_N$; and if they are independent, we should be able to solve for a unique set of node voltages.

It is usually convenient to rearrange the equations so that the coefficients of V_i are in a column, and the current source terms are on the right-hand side:

V_1	V_2	V_3	V_4	V_5	
$G_{10}+G_{12}$	$-G_{12}$	0	0	0	$= -I12$
$-G_{12}$	$G_{12}+G_{23}$	$-G_{23}$	0	0	$= I12$
0	$-G_{23}$	$G_{23}+G_{30}+$ $G_{35}+G_{34}$	$-G_{34}$	$-G_{35}$	$= I03$
0	0	$-G_{34}$	$G_{34}+G_{45}$	$-G_{45}$	$= I54$
0	0	$-G_{35}$	$-G_{45}$	$G_{50}+G_{35}+G_{45}$	$= -I54$

We have used the conductance $G_{ij} = 1/R_{ij}$ for convenience. Notice that the array of coefficients could have been written by inspection from the network. There is a positive diagonal term in the ith row corresponding to each conductance attached to node i; and there is a negative off-diagonal term in row i and column j corresponding to a conductance from node j to node i. Finally, the right-hand side of equation i is simply the net strength of current sources into node i. We next describe an algorithm for solving these equations on a digital computer.

Exercise 11.3.1

Show that the equation obtained from KCL at node "0" is not independent of the set of N equations written at nodes 1 through N.

Exercise 11.3.2

Write the coefficient array and right-hand side of the node equations for the network in Fig. 11.3.3 ($\mho$ is the symbol for mho, the unit of conductance).

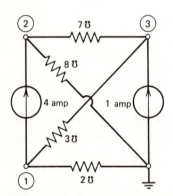

FIG. 11.3.3   Resistor network for Exercise 11.3.2.

11.4 THE GAUSSIAN ELIMINATION ALGORITHM

To illustrate the Gaussian elimination method of solving simultaneous linear equations, we shall use the example in Fig. 11.4.1.

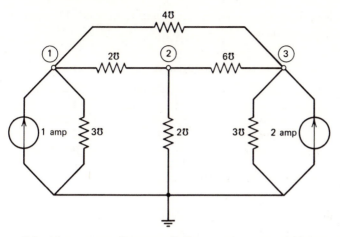

FIG. 11.4.1   Network used to illustrate the Gaussian elimination method.

The coefficient array and right-hand side become

V_1	V_2	V_3	
9	−2	−4	=1
−2	10	−6	=0
−4	−6	13	=2

The idea is to solve the first equation for V_1 in terms of V_2 and V_3, and to substitute this into equations 2 and 3. Thus

$$V_1 = \tfrac{1}{9}(1 + 2V_2 + 4V_3)$$

The set of equations then becomes

$$9V_1 + (-2)V_2 + (-4)V_3 = 1$$

$$0V_1 + \left(10 - \frac{4}{9}\right)V_2 + \left(-6 - \frac{8}{9}\right)V_3 = \frac{2}{9}$$

$$0V_1 + \left(-6 - \frac{8}{9}\right)V_2 + \left(13 - \frac{16}{9}\right)V_3 = 2 + \frac{4}{9}$$

Notice that the second equation can be obtained by dividing the first equation by 9, the coefficient of V_1; multiplying it by −2, the coefficient of V_1 in equation 2; and subtracting the result from equation 2. Similarly, the third equation can be obtained by multiplying the first equation by $(-4/9)$ and subtracting the result from equation 3. The coefficient 9 is called a *pivot*, and the operation of eliminating variable i below the ith equation is called *pivoting*. Rewriting the coefficient array and right-hand side after pivoting to eliminate V_1, we have:

V_1	V_2	V_3	
9	−2	−4	=1
0	9.556	−6.889	= .222
0	−6.889	11.222	=2.444

Continuing by eliminating V_2 from equation 3, we get

V_1	V_2	V_3	
9	−2	−4	=1
0	9.556	−6.889	= .222
0	0	6.256	=2.605

We can now solve equation 3 for V_3. Knowing V_3, we can solve equation 2 for V_2. Finally, knowing V_2 and V_3, we can solve equation 1 for V_1. This process is called *back substitution*. Thus, we obtain

$$V_3 = .416 \text{ volts}$$
$$V_2 = .323 \text{ volts}$$
$$V_1 = .368 \text{ volts}$$

Exercise 11.4.1

Solve the equations obtained in Exercise 11.3.2.

11.5 A COMPUTER PROGRAM
FOR GAUSSIAN ELIMINATION

The Gaussian elimination procedure can be programmed in a straightforward way. Figure 11.5.1 shows a program in subroutine form, with the coefficient array A(I,J), the right-hand side C(I), the unknowns B(I), and N all in a COMMON block called BLOCK1. The only subtlety in the program involves the choice of a pivoting element. In order to eliminate the Ith unknown, that equation among equations I to N is chosen whose leading term has the largest absolute value. This is done in the DO 11 loop.

If the pivot found is very small (less than 10^{-10} in absolute value) the program branches to statement 13, where the message "ill-conditioned" is printed, and control is returned to the calling program. This test prevents attempted division by zero in case the equations are not independent, and the choice of pivot minimizes the effect of round-off error.

The DO 14 loop then switches rows so that the largest pivot occurs in row I. The two nested DO 15 loops then perform the pivoting operation. The last pivot, A(N,N), is then tested; and the DO 17 loop then performs the back substitution.

```
      SUBROUTINE GAUSS
      COMMON/BLOCK1/A(15,15),B(15),C(15),N
C.....ELIMINATE I-TH UNKNOWN
      NM=N-1
      DO 15 I=1,NM
C.....FIND LARGEST PIVOT
      AMAX=0.
      DO 11 II=I,N
      IF(ABS(A(II,I)).LT.AMAX)GOTO11
      ISTAR=II
      AMAX=ABS(A(II,I))
   11 CONTINUE
C.....RETURN IF PIVOT IS TOO SMALL
      IF(AMAX.LT.1.E-10)GOTO13
C.....SWITCH ROWS
      DO 14 J=I,N
      DUM=A(ISTAR,J)
      A(ISTAR,J)=A(I,J)
   14 A(I,J)=DUM
      DUM=C(ISTAR)
      C(ISTAR)=C(I)
      C(I)=DUM
C.....PIVOT
      IP=I+1
      DO 15 II=IP,N
      PIVOT=A(II,I)/A(I,I)
      C(II)=C(II)-PIVOT*C(I)
      DO 15 J=I,N
   15 A(II,J)=A(II,J)-PIVOT*A(I,J)
C.....RETURN IF LAST PIVOT IS TOO SMALL
      IF(ABS(A(N,N)).LT.1.E-10)GOTO13
      B(N)=C(N)/A(N,N)
C.....BACK SUBSTITUTE
      DO 17 K=1,NM
      L=N-K
      B(L)=C(L)
      LP=L+1
      DO 16 J=LP,N
   16 B(L)=B(L)-A(L,J)*B(J)
   17 B(L)=B(L)/A(L,L)
      RETURN
   13 WRITE(6,12)
   12 FORMAT(' ILL-CONDITIONED')
      RETURN
      END
```

FIG. 11.5.1 FORTRAN program in the form of SUBROUTINE GAUSS, for performing Gaussian elimination.

11.6 A COMPLETE PROGRAM FOR SOLVING RESISTOR NETWORKS

We shall now write a program that will read in a description of a resistor network with current sources and set up the appropriate linear equations so that SUBROUTINE GAUSS can be used to solve for the resulting voltage distribution. Resistor networks are, of course, graphs, and any of the methods discussed before can be used for their description. Actually, there are two graphs to describe: one consisting of the conductances and one consisting of the current sources. We shall use two

```
C......THIS PROGRAM SETS UP THE EQUATIONS DESCRIBING A RESISTOR NETWORK
       COMMON/BLOCK1/A(15,15),B(15),C(15),N
       INTEGER N1G(100),N2G(100),N1S(100),N2S(100)
       REAL G(100),S(100)
C......READ AND WRITE # OF NODES, CONDUCTANCES, AND SOURCES; IN BR.-LIST
C......FORM - NOTE THAT N DOES NOT INCLUDE REFERENCE NODE # 0
       READ(5,1)N
     1 FORMAT(I2)
       WRITE(6,2)N
     2 FORMAT('1# OF NODES=',I5)
       READ(5,3)NG,(N1G(J),N2G(J),G(J),J=1,NG)
     3 FORMAT(I2/(2I2,F10.4))
       WRITE(6,4)NG,(N1G(J),N2G(J),G(J),J=1,NG)
     4 FORMAT(' # OF ELEMENTS=',I5/(2I4,F10.4))
       READ(5,3)NS,(N1S(J),N2S(J),S(J),J=1,NS)
       WRITE(6,4)NS,(N1S(J),N2S(J),S(J),J=1,NS)
C......INITIALIZE A,B,C
       DO 5 I=1,N
       B(I)=0.
       C(I)=0.
       DO 5 J=1,N
     5 A(I,J)=0.
C......PUT CONDUCTANCES IN A AND SOURCES IN C
       DO 7 I=1,NG
       L1=N1G(I)
       L2=N2G(I)
       IF(L1.EQ.0.OR.L2.EQ.0)GOTO6
       A(L1,L2)=A(L1,L2)-G(I)
       A(L2,L1)=A(L2,L1)-G(I)
     6 IF(L1.NE.0)A(L1,L1)=A(L1,L1)+G(I)
       IF(L2.NE.0)A(L2,L2)=A(L2,L2)+G(I)
     7 CONTINUE
     - DO 8 I=1,NS
       L1=N1S(I)
       L2=N2S(I)
       IF(L1.NE.0)C(L1)=C(L1)-S(I)
       IF(L2.NE.0)C(L2)=C(L2)+S(I)
     8 CONTINUE
C......WRITE MATRIX A AND VECTOR C
       DO 9 I=1,N
     9 WRITE(6,10)(A(I,J),J=1,N),C(I)
    10 FORMAT(10(' ',F10.4))
C......SOLVE LINEAR EQUATIONS AND WRITE ANSWER
       CALL GAUSS
       WRITE(6,11)(I,B(I),I=1,N)
    11 FORMAT(' B(',I2,')=',E16.8)
       STOP
       END
```

FIG. 11.6.1 A complete FORTRAN program for finding the voltages in a resistor network. This program calls SUBROUTINE GAUSS to solve simultaneous equations.

branch-lists: In the first list, the entry N1G(I),N2G(I) will represent the Ith conductance, which will go from node N1G(I) to N2G(I), and which has the value G(I) mhos. Altogether, there will be NG conductances. In the second list, the entry N1S(I),N2S(I) will represent the Ith current source, with strength S(I), and going from node N1S(I) to N2S(I). Thus, the conductance branch-list represents an undirected graph, since the conductances do not have any directions associated with them, while the current source branch-list represents a directed graph, since the sources do have directions associated with them.

Figure 11.6.1 shows the computer program that sets up the simultaneous equations describing the resistor network. This is accomplished by filling in the matrix A and the vector C, and then calling SUBROUTINE GAUSS to solve the equations. After the usual storage allocation declarations, the program begins by reading and writing the number of nodes, N; the number of conductances, NG; the conductance branch-list, N1G(I),N2G(I); the values of the conductances, G(I); the number of sources, NS; the source branch-list, N1S(I),N2S(I); and the values of the sources, S(I). Next, with the DO 5 loops, the arrays A,B, and C are initialized to zero. The DO 7 loop fills up the A matrix according to the rules discussed in Section 11.3, and similarly the DO 8 loop fills up the C vector using information about the current sources. The A matrix and C vector are then written out, SUBROUTINE GAUSS called to solve the equations, and the solution printed out. Figure 11.6.2 shows the output when data describing the network of Fig. 11.4.1 is supplied to the program. Note that N = 3 in this case; that is, the ground, or reference, node is not counted as a node.

```
# OF NODES=      3
# OF ELEMENTS=      6
   0    1     3.0000
   0    2     2.0000
   0    3     3.0000
   1    2     2.0000
   2    3     6.0000
   1    3     4.0000
# OF ELEMENTS=      2
   0    1     1.0000
   0    3     2.0000
     9.0000      -2.0000      -4.0000      1.0000
    -2.0000      10.0000      -6.0000      0.0000
    -4.0000      -6.0000      13.0000      2.0000
B( 1)=    0.36802930E 00
B( 2)=    0.32341950E 00
B( 3)=    0.41635650E 00
```

FIG. 11.6.2 The output of the program shown in the previous figure when run on the network in Fig. 11.4.1.

Exercise 11.6.1

Will the program for finding voltages in a resistor network with current sources work if N=1? If not, suggest a change that will allow the program to operate properly in this case.

11.7 THE PHASOR RESPONSE OF RLC NETWORKS

The remainder of this chapter will be devoted to extending what we have done for resistor networks to networks that might include capacitors and inductors as well. In many respects our analysis will be analogous to the phasor analysis of moving average filters in Chapter 3; that is, we shall investigate the response of RLC networks to current sources that vary sinusoidally as functions of time. The solutions for voltages in the network will be phasors, and will represent the sinusoidal steady state response of the network to a phasor at a fixed frequency.

To be more concrete, consider the network in Fig. 11.7.1 where we have indicated a capacitor and an inductor by the conventional symbols. (It is assumed here that the reader is familiar with the mathematical models for the linear constant capacitor and inductor. If not, he should consult an introductory text on network theory, such as those in the suggestions for further reading at the end of this chapter.) In this network, the current source is denoted by $i(t)$, since we are now interested in the time variation of currents and voltages. This is in contrast with the resistor network situation, where the strength of a current source was assumed to be a constant function of time, and all the voltages and currents were, in fact, constant.

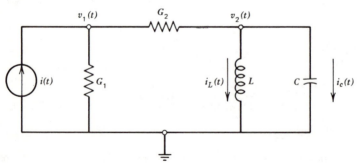

FIG. 11.7.1 RLC network.

Similarly, we denote the voltages at nodes 1 and 2 by $v_1(t)$ and $v_2(t)$, since these will, in general, vary with time t. For every time t we can write KCL at the two nodes as follows, where, for the moment, we have denoted the current leaving node 2 through the capacitor by i_C and the current leaving node 2 through the inductor by i_L:

$$G_1 v_1 + (v_1 - v_2) G_2 = i$$
$$(v_2 - v_1) G_2 + i_C + i_L = 0 \qquad (11.7.1)$$

We can express i_C in terms of the voltage at node 2 by using the mathematical model for a linear constant capacitor:

$$i_C = C \left(\frac{dv_2}{dt} \right) \qquad (11.7.2)$$

Similarly, we can express the current through the inductor in terms of the voltage across it as follows:

$$i_L = \left(\frac{1}{L} \right) \int v_2 \, dt + \text{constant} \qquad (11.7.3)$$

If we substitute these expressions for i_C and i_L into Eq. 11.7.1, we obtain two simultaneous equations in v_1 and v_2, analogous to those obtained for resistor networks in Section 10.3.

$$G_1 v_1 + (v_1 - v_2) G_2 = i$$
$$(v_2 - v_1) G_2 + C \left(\frac{dv_2}{dt} \right) + \left(\frac{1}{L} \right) \int v_2 dt + \text{constant} = 0 \qquad (11.7.4)$$

The marked difference is that now the equations contain the derivative and integral of voltage, as well as the voltage variables themselves. We call a set of equations like (11.7.4) "simultaneous integro-differential equations." They cannot be solved in general terms simply by applying algebraic operations; that is, Gauss elimination does not work.

If, however, we assume that the current source i is a phasor, we can reduce the solution of these equations to algebraic steps. By a phasor in this case we mean a continuous-time phasor such as those discussed in Section 2.4:

$$i(t) = I_{01} e^{j\omega t}$$

Here I_{01} is a complex number that represents the amplitude and phase angle of the phasor signal $i(t)$, and ω represents the frequency, in radians/second. In contrast with the discrete-time case, there is no restriction on how large ω may be. If we substitute this expression for $i(t)$ in (11.7.4) we obtain

$$G_1 v_1 + (v_1 - v_2) G_2 = I_{01} e^{j\omega t}$$
$$(v_2 - v_1) G_2 + C \left(\frac{dv_2}{dt} \right) + \left(\frac{1}{L} \right) \int v_2 \, dt + \text{constant} = 0 \qquad (11.7.5)$$

Recall that for a digital filter with a phasor input, the output was a phasor of the same frequency. This situation is very closely analogous to the digital filter situation: the solutions of Eq. 11.7.5 will be phasor voltages, with amplitudes and phases that are to be determined. To verify this statement, let us propose as solutions to Eq. 11.7.5 the following phasor voltages:

$$v_1 = V_1 e^{j\omega t}$$
$$v_2 = V_2 e^{j\omega t}$$

(11.7.6)

The term involving the derivative of v_2 will become

$$\frac{dv_2}{dt} = (j\omega) V_2 e^{j\omega t}$$

(11.7.7)

and the term involving the integral of v_2 will become

$$\int v_2 dt = \left(\frac{1}{j\omega}\right) V_2 e^{j\omega t}$$

(11.7.8)

Equations 11.7.5 then become

$$G_1 V_1 e^{j\omega t} + (V_1 e^{j\omega t} - V_2 e^{j\omega t}) G_2 = I_{01} e^{j\omega t}$$
$$(V_2 e^{j\omega t} - V_1 e^{j\omega t}) G_2 + (j\omega C) V_2 e^{j\omega t} + \left(\frac{1}{j\omega L}\right) V_2 e^{j\omega t} + \text{constant} = 0$$

(11.7.9)

Notice now that if we take the constant to be zero, every remaining term in these equations has associated with it the factor $e^{j\omega t}$. The constant is associated with the inductor, and represents a constant, or zero frequency, component of the current through the inductor. In the networks we shall consider, these constants can always be taken to be zero. If we do this, and if we divide every equation by $e^{j\omega t}$, we arrive at the following set of simultaneous equations in the unknowns V_1 and V_2:

$$G_1 V_1 + (V_1 - V_2) G_2 = I_{01}$$
$$(V_2 - V_1) G_2 + (j\omega C) V_2 + \left(\frac{1}{j\omega L}\right) V_2 = 0$$

(11.7.10)

A number of observations about these equations are in order. First, they are simultaneous *algebraic* equations in the unknown voltage variables. We have thus transformed the original set of differential equations into algebraic equations by assuming that the current sources are phasors. This is completely analogous to what we did for moving average digital filters, where we transformed a difference equation into an algebraic equation in the same way.

A second observation concerns the fact that these simultaneous equations have coefficients that are possibly complex numbers, and that the unknown voltage variables can take on complex values in the solution.

The method of Gauss elimination applies equally well to complex numbers, however, so that this additional complication will cause no extra difficulty in obtaining the solution of these equations.

A last observation concerns the fact that these equations can be written by inspection from the network in much the same way as for a resistor network. The additional rules concern the treatment of capacitors and inductors and are summarized as follows:

(a) A capacitor is treated in exactly the same way as a conductance, except that the coefficient in the equations is $(j\omega C)$.

(b) An inductor is treated in exactly the same way as a conductance, except that the coefficient in the equations is $1/(j\omega L)$.

Thus the coefficients of the linear simultaneous equations depend on the frequency of the current sources, ω. This means that if we use Gauss elimination to solve the equations, we must re-solve the equations for each frequency of interest.

Returning to our example, the equations describing the network of Fig. 11.7.1 can now be written in schematic form as:

V_1	V_2	
$G_1 + G_2$	$-G_2$	$= I_{01}$
$-G_2$	$G_2 + j\omega C + 1/(j\omega L)$	$= 0$

$$(11.7.11)$$

For each value of the frequency ω, we can solve these equations for V_1 and V_2 using Gauss elimination. In the next section, we shall describe computer programs for Gauss elimination for complex numbers, and for setting up the equations describing general RLC networks.

It is of interest to note that in this simple example it is quite convenient to solve Eqs. 11.7.11 analytically for V_2, say, in terms of the current source strength I_{01}. If we solve the first equation for V_1 in terms of V_2 and substitute into the second equation, we can solve for V_2 in terms of the current source strength I_{01}:

$$V_2 = \left[\frac{G_2/(G_1 + G_2)}{G_1 G_2/(G_1 + G_2) + j\omega C + 1/(j\omega L)} \right] \cdot I_{01}$$

$$(11.7.12)$$

The quantity in brackets can be thought of as a transfer function, in much the same way as the transfer function of a digital filter: it represents the effect on the magnitude and phase of the phasor "input" current source of the network, if we consider V_2 the "output" of the system.

Take the case where $G_1 = G_2 = 1$ mho, $C = 1$ farad, and $L = 1$ henry; then we can write the transfer function numerically as

$$H(\omega) = \frac{.5}{.5 + j\omega - j/\omega} \qquad (11.7.13)$$

The magnitude of this transfer function can be written as a function of the frequency ω as

$$|H(\omega)| = \frac{.5}{\sqrt{.25 + (\omega - 1/\omega)^2}} \qquad (11.7.14)$$

and we see that the magnitude of the transfer function has a peak at the frequency $\omega = 1$ radian/second. This is analogous to the peak in the transfer function of a two-pole resonant digital filter, such as those discussed in Section 5.2.

Just as the phasor approach can be extended by the introduction of the z-transform in the digital case, the introduction of the Laplace transform will allow a general treatment of analog circuits. This brings us to the subject of linear circuit theory, which it is not our purpose to discuss here, and which we leave for a later course. We should mention, however, that many of the ideas developed for digital filters carry over to the analog case, especially those involving the complex frequency plane, poles, zeros, and frequency response. In addition, the graph theory ideas prove useful in studying the general properties of electrical networks in greater detail than we have done here.

Exercise 11.7.1

Set up the complex equations describing the network shown in Fig. 11.7.2.

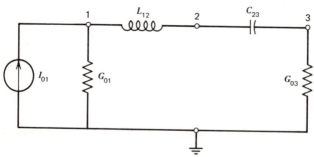

FIG. 11.7.2   RLC network for Exercise 11.7.1.

Exercise 11.7.2

Repeat for the networks in Fig. 11.7.3.

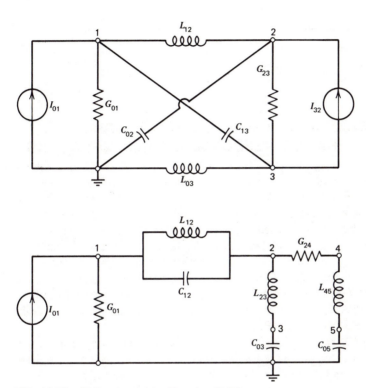

FIG. 11.7.3 RLC networks for Exercise 11.7.2.

Exercise 11.7.3

(a) Write Kirchhoff's current law at node 1 for the network shown in Fig. 11.7.4.

(b) Let $i(t)$ be the continuous-time phasor $Ie^{j\omega t}$, and assume $v(t)$ is the continuous-time phasor $Ve^{j\omega t}$. Solve for the transfer function V/I as a function of frequency ω.

(c) Sketch the magnitude of the transfer function as a function of the frequency ω.

(d) Is this transfer function low-pass or high-pass?

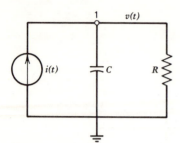

FIG. 11.7.4 Network for Exercise 11.7.3.

11.8 A COMPUTER PROGRAM FOR FINDING THE PHASOR RESPONSE OF RLC NETWORKS

As mentioned above, the simultaneous equations describing the phasor response of an RLC network can be solved using Gauss elimination, and Fig. 11.8.1 shows an appropriate Gauss elimination subroutine. This program is identical to that used for resistor networks, except that it now accepts complex elements in the matrix A, the vector C, and the solution vector B, and the arithmetic operations carried out use complex arithmetic. The only differences in the FORTRAN code between this program and the one shown in Fig. 11.5.1 are that this program declares the variables A,B,C, DUM, and PIVOT to be complex; and that this program uses the built-in function CABS in place of ABS.

Figure 11.8.2 shows a program analogous to the one already described for setting up the equations describing resistor networks. The new program begins by reading in the information describing the network: the number of nodes, the node at which the voltage will be printed, the number of conductances, the values of the conductances, the number of capacitances, their values, and so on. Note that the coefficients of the current sources are read in as real numbers, so that the phase associated with all the current sources is assumed to be zero. The matrix A and the vector C are set up using the rules discussed above, and SUBROUTINE GAUSS is called to solve for the voltages. The magnitude and phase of the voltage of interest is then printed out, and this process is repeated for 50 values of the frequency, by virtue of the DO 20 loop. Fig. 11.8.3 shows the output when the network of Fig. 11.7.1 is run, and the voltage V(2) is specified as the voltage of interest. Figure 11.8.4 shows a plot of the magnitude of V(2) versus the frequency ω, as ω varies from .2 to 5 radians/second. Notice that the curve has the predicted peak at the frequency $\omega = 1$ radian/second.

```
      SUBROUTINE GAUSS
      COMMON/BLOCK1/A(15,15),B(15),C(15),N
      COMPLEX A,B,C,DUM,PIVOT
C......ELIMINATE I-TH UNKNOWN
      NM=N-1
      DO 15 I=1,NM
C......FIND LARGEST PIVOT
      AMAX=0.
      DO 11 II=I,N
      IF(CABS(A(II,I)).LT.AMAX)GOTO11
      ISTAR=II
      AMAX=CABS(A(II,I))
   11 CONTINUE
C......RETURN IF PIVOT IS TOO SMALL
      IF(AMAX.LT.1.E-10)GOTO13
C......SWITCH ROWS
      DO 14 J=I,N
      DUM=A(ISTAR,J)
      A(ISTAR,J)=A(I,J)
   14 A(I,J)=DUM
      DUM=C(ISTAR)
      C(ISTAR)=C(I)
      C(I)=DUM
C......PIVOT
      IP=I+1
      DO 15 II=IP,N
      PIVOT=A(II,I)/A(I,I)
      C(II)=C(II)-PIVOT*C(I)
      DO 15 J=I,N
   15 A(II,J)=A(II,J)-PIVOT*A(I,J)
C......RETURN IF LAST PIVOT IS TOO SMALL
      IF(CABS(A(N,N)).LT.1.E-10)GOTO13
      B(N)=C(N)/A(N,N)
C......BACK SUBSTITUTE
      DO 17 K=1,NM
      L=N-K
      B(L)=C(L)
      LP=L+1
      DO 16 J=LP,N
   16 B(L)=B(L)-A(L,J)*B(J)
   17 B(L)=B(L)/A(L,L)
      RETURN
   13 WRITE(6,12)
   12 FORMAT(' ILL-CONDITIONED')
      RETURN
      END
```

FIG. 11.8.1 FORTRAN program in the form of SUBROUTINE GAUSS, for performing Gaussian elimination with complex numbers.

```
C......THIS PROGRAM SETS UP AND SOLVES EQUATIONS DESCRIBING RLC NETWORK
       COMMON/BLOCK1/A(15,15),B(15),C(15),N
       COMPLEX A,B,C,JAY,TERM,CMPLX
       INTEGER N1G(100),N2G(100),N1CAP(100),N2CAP(100),N1L(100),N2L(100),
     X        N1S(100),N2S(100)
       REAL G(100),CAP(100),L(100),S(100)
       JAY=CMPLX(0.,1.)
       PI=3.14159265
C......READ AND WRITE # OF NODES, CONDUCTANCES, CAPACITANCES,
C......INDUCTANCES, AND SOURCES, IN BR.-LIST FORM
C......NOTE THAT N DOES NOT INCLUDE THE REFERENCE NODE # 0
C......ISTAR IS NODE WHOSE VOLTAGE IS DESIRED
       READ(5,1)N,ISTAR
     1 FORMAT(2I2)
       WRITE(6,2)N,ISTAR
     2 FORMAT('1# OF NODES=',I3,' CALCULATING VOLTAGE AT NODE',I3)
       READ(5,3)NG,(N1G(J),N2G(J),G(J),J=1,NG)
     3 FORMAT(I2/(2I2,F10.4))
       WRITE(6,4)NG,(N1G(J),N2G(J),G(J),J=1,NG)
     4 FORMAT(' # OF ELEMENTS=',I5/(2I4,F10.4))
       READ(5,3)NCAP,(N1CAP(J),N2CAP(J),CAP(J),J=1,NCAP)
       WRITE(6,4)NCAP,(N1CAP(J),N2CAP(J),CAP(J),J=1,NCAP)
       READ(5,3)NL,(N1L(J),N2L(J),L(J),J=1,NL)
       WRITE(6,4)NL,(N1L(J),N2L(J),L(J),J=1,NL)
       READ(5,3)NS,(N1S(J),N2S(J),S(J),J=1,NS)
       WRITE(6,4)NS,(N1S(J),N2S(J),S(J),J=1,NS)
C......DO FOR 50 FREQUENCY POINTS
       DO 20 NFREQ=1,50
       W=.2*FLOAT(NFREQ)
C......INITIALIZE A,B,C
       DO 5 I=1,N
       B(I)=0.
       C(I)=0.
       DO 5 J=1,N
     5 A(I,J)=0.
C......PUT CONDS., CAPS., AND INDS. IN A AND SOURCES IN C
       DO 7 I=1,NG
       L1=N1G(I)
       L2=N2G(I)
       IF(L1.EQ.0.OR.L2.EQ.0)GOTO6
       A(L1,L2)=A(L1,L2)-G(I)
       A(L2,L1)=A(L2,L1)-G(I)
     6 IF(L1.NE.0)A(L1,L1)=A(L1,L1)+G(I)
       IF(L2.NE.0)A(L2,L2)=A(L2,L2)+G(I)
     7 CONTINUE
       DO 9 I=1,NCAP
       L1=N1CAP(I)
       L2=N2CAP(I)
       TERM=JAY*W*CAP(I)
       IF(L1.EQ.0.OR.L2.EQ.0)GOTO8
       A(L1,L2)=A(L1,L2)-TERM
       A(L2,L1)=A(L2,L1)-TERM
     8 IF(L1.NE.0)A(L1,L1)=A(L1,L1)+TERM
       IF(L2.NE.0)A(L2,L2)=A(L2,L2)+TERM
     9 CONTINUE
       DO 11 I=1,NL
       L1=N1L(I)
       L2=N2L(I)
       TERM=1./(JAY*W*L(I))
```

FIG. 11.8.2   A complete FORTRAN program for finding the phasor voltages in an RLC network (continued on next page).

```
      IF(L1.EQ.0.OR.L2.EQ.C)GOTO10
      A(L1,L2)=A(L1,L2)-TERM
      A(L2,L1)=A(L2,L1)-TERM
   10 IF(L1.NE.0)A(L1,L1)=A(L1,L1)+TERM
      IF(L2.NE.0)A(L2,L2)=A(L2,L2)+TERM
   11 CONTINUE
      DO 12 I=1,NS
      L1=N1S(I)
      L2=N2S(I)
      IF(L1.NE.0)C(L1)=C(L1)-S(I)
      IF(L2.NE.0)C(L2)=C(L2)+S(I)
   12 CONTINUE
C......SOLVE LINEAR EQUATICNS AND WRITE ANSWER
      CALL GAUSS
      BR=REAL(B(ISTAR))
      BI=AIMAG(B(ISTAR))
      BM=SQRT(BR*BR+BI*BI)
      BP=ATAN2(BI,BR)*180./PI
   20 WRITE(6,14)W,BM,BP
   14 FORMAT(' W=',E16.8,' BM=',E16.8,' BP=',E16.8)
      STOP
      END
```

Exercise 11.8.1

Solve for the transfer function V_3/I_{01} in the network of Fig. 11.7.2.

Exercise 11.8.2

Show that the solution for any voltage in an RLC network with phasor current sources is a rational function of $(j\omega)$.

Exercise 11.8.3

Consider an RLC network with n current sources with strengths $I_1, I_2, \ldots, I_n$. Show that the solution for a voltage V is a linear function of $I_1, I_2, \ldots, I_n$.

```
#  CF NODES=  2 CALCULATING VOLTAGE AT NODE  2
#  OF ELEMENTS=    2
   0    1    1.0000
   1    2    1.0000
#  OF ELEMENTS=    1
   0    2    1.0000
#  OF ELEMENTS=    1
   0    2    1.0000
#  OF ELEMENTS=    1
   0    1    1.0000
W=  0.1999999CE 00  BM=  0.10360610E 00  BP=  0.84053080E 02
W=  0.39999990E 00  BM=  0.23162040E 00  BP=  C.76607460E 02
W=  0.59999990E 00  FM=  0.42443410E 00  BP=  0.64885100E 02
W=  0.79999990E 00  BM=  0.7432944CE 00  BP=  0.41987150E 02
W=  0.99999990E 00  BM=  0.10000000E 01  BP=  C.00000000E 00
W=  0.11999990E 01  BM=  0.8064C500E 00  BP= -C.36253780E 02
W=  0.13999990E 01  BM=  0.58917240E 00  BP= -0.539C1680E 02
W=  0.15999990E 01  BM=  0.4563170CE 00  BP= -0.62850250E 02
W=  0.17999990E 01  BM=  0.3728190CE 00  BP= -0.68110380E 02
W=  0.19999990E 01  BM=  0.31622780E 00  BP= -0.71564970E 02
W=  0.21999990F 01  BM=  0.27538230E 00  BP= -C.74015150E 02
W=  0.23999990E 01  BM=  0.24445240E 00  BP= -0.75850500E 02
W=  0.25999990E 01  BM=  0.2201568CE 00  BP= -0.772E1730F 02
W=  0.27999990E 01  BM=  0.20052110E 00  BP= -0.78432540E 02
W=  0.29999990E 01  BM=  0.1842884CE 00  BP= -0.79380290E 02
W=  0.31999990E 01  BM=  0.17062100E 00  BP= -0.80176040E 02
W=  0.33999990E 01  BM=  0.1589384CE 00  BP= -C.80854690E 02
W=  0.35999990E 01  BM=  0.14882550E 00  BP= -0.81441100E 02
W=  0.37999990E 01  BM=  0.13997710E 00  BP= -0.81953380E 02
W=  0.39999990E 01  BM=  0.13216360E 00  BP= -0.82405270E 02
W=  0.41999980E 01  BM=  0.12520870E 00  BP= -0.82807120E 02
W=  0.43999990E 01  BM=  0.11897460E 00  BP= -0.83167020E 02
W=  0.45999990E 01  BM=  0.11335190E 00  BP= -0.83491390E 02
W=  0.47999990E 01  PM=  0.10825290E 00  BP= -0.83785290E 02
W=  0.49999990E 01  BM=  0.10360610E 00  BP= -0.84053C80E 02
W=  0.51999980E 01  BM=  0.99352350E-01  BP= -C.84298030E 02
W=  0.53999990E 01  BM=  0.95442950E-01  BP= -0.84523100E 02
W=  0.55999990E 01  BM=  0.91836750E-01  BP= -0.84730660E 02
W=  0.57999990E 01  BM=  0.88499360E-01  BP= -0.84922690E 02
W=  0.59999990E 01  BM=  0.85401110E-01  BP= -0.851CC890E 02
W=  0.61999980E 01  BM=  0.82516720E-01  BP= -0.85266640E 02
W=  0.63999990E 01  BM=  0.798245C0E-01  BP= -C.85421440E 02
W=  0.65999990E 01  BM=  0.77305490E-01  BP= -0.85566220E 02
W=  0.67999990E 01  BM=  0.74943300E-01  BP= -0.85701990E 02
W=  0.69999990E 01  BM=  0.7272350CE-01  BP= -C.85829510E 02
W=  0.71999980E 01  BM=  0.70633470E-01  BP= -0.85949500E 02
W=  0.73999990E 01  BM=  0.6866192CE-01  BP= -0.86062800E 02
W=  0.75999990E 01  BM=  0.66798920E-01  BP= -0.86169810E 02
W=  0.77999990E 01  BM=  0.6503576CE-01  BP= -C.86271080E 02
W=  0.79999990E 01  BM=  0.63364380E-01  BP= -0.86366970E 02
W=  0.81999980E 01  BM=  0.61777910E-01  BP= -0.86458050E 02
W=  0.83999980E 01  BM=  0.60269780E-01  BP= -C.86544630E 02
W=  0.85999990E 01  BM=  0.58834300E-01  BP= -0.86626990E 02
W=  0.87999990E 01  BM=  0.57466360E-01  BP= -C.86705560E 02
W=  0.89999990F 01  BM=  0.56161190E-01  BP= -0.86780480E 02
W=  0.91999980E 01  BM=  0.54914590E-01  BP= -0.86851970E 02
W=  0.93999990E 01  BM=  0.53722660E-01  BP= -0.86920360E 02
W=  0.95999990E 01  BM=  0.5258182CE-01  BP= -0.86985800E 02
W=  0.97999990E 01  BM=  0.51488820E-01  BP= -C.87048520E 02
W=  0.99999990E 01  BM=  0.50440750E-01  BP= -C.87108650E 02
```

FIG. 11.8.3 The output of the program shown in the previous figure when run on the network in Fig. 11.7.1.

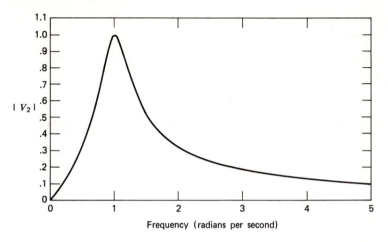

FIG. 11.8.4   The magnitude of the phasor voltage V_2 vs. frequency in the example.

Further Reading

The material in this chapter leads naturally to a traditional electrical engineering course in circuit theory, for which there are many excellent texts. The following list is representative, but by no means exhaustive.

1. *Network Analysis,* M. E. Van Valkenburg, Prentice-Hall, Englewood Cliffs, N.J., 1964, Second Edition.

2. *Network Theory, An Introductory Course,* T. S. Hwang and R. R. Parker, Addison-Wesley, Reading, Mass., 1971.

3. *Basic Network Theory,* P. M. Chirlian, McGraw-Hill, New York, 1969.

4. *Circuit Theory; An Introduction to the State Variable Approach,* R. A. Rohrer, McGraw-Hill, New York, 1970.

5. *Circuits, Devices, and Systems,* R. J. Smith, Wiley, New York, 1971, Second Edition.

6. *Circuits, Signals, and Networks,* C. W. Cox and W. L. Reuter, Macmillan, New York, 1969.

7. *Introductory Topological Analysis of Electrical Networks,* S. P. Chan, Holt, Rinehart and Winston, New York, 1969.

8. *Electrical Networks,* L. O. Chua, McGraw-Hill, New York, 1969.

9. *Basic Circuit Theory,* C. A. Desoer and E. S. Kuh, McGraw-Hill, New York, 1969.

10. *Modern Network Analysis,* W. H. Kim and H. E. Meadows, Jr., Wiley, New York, 1971.

11. *Linear Networks and Systems,* B. C. Kuo, McGraw-Hill, New York, 1967.

12. *Introductory Signals and Circuits,* J. B. Cruz, Jr. and M. E. Van Valkenburg, Blaisdell, Waltham, Mass., 1967.

APPENDIX

ANSWERS AND HINTS TO SELECTED EXERCISES

CHAPTER 1

1.2.2 The nth root of $z = Re^{j\theta}$ is not unique; there are n of them given by

$$R^{1/n}e^{j(\theta+2\pi k)/n}, \, k=0,\ldots,n-1.$$

The nth power of $z = Re^{j\theta}$ is $R^n \, e^{jn\theta}$

1.2.3 If $z = Re^{j\theta}$, $z^* = Re^{-j\theta}$.

$$\text{Real}\,(z) = \tfrac{1}{2}\,(z+z^*)$$
$$\text{Imag}\,(z) = \frac{1}{2j}\,(z-z^*)$$
$$|z| = \sqrt{zz^*}.$$

1.2.5 *Hint:* use induction.

1.2.7 No.

1.2.9 Calculate in order:

$q = a+b$	1 addition
$r = c-d$	1 negation, 1 addition
$s = qr$	1 multiplication
$t = bc$	1 multiplication
$u = -t$	1 negation
$v = ad$	1 multiplication
$\text{Real}\,(wz) = s+u+v$	2 additions
$\text{Imag}\,(wz) = v+t$	1 addition.

1.2.13

$$\cos n\theta = \sum_{\substack{k=0 \\ k \text{ even}}}^{n} (-1)^{k/2} \binom{n}{k} (\cos \theta)^{n-k} (\sin \theta)^{k}$$

1.3.1

$$\cos \theta = \tfrac{1}{2}(e^{j\theta} + e^{-j\theta})$$

$$\sin \theta = \frac{1}{2j}(e^{j\theta} - e^{-j\theta}).$$

1.3.2

$$|F| = \sqrt{2 + 2\cos \theta}$$

$$\arg F = \arctan \frac{\sin \theta}{1 + \cos \theta}.$$

The locus of F is a circle of radius 1 centered at $z = 1$.

1.3.3 If $z = jx$,

$$|F| = \frac{1}{\sqrt{1 + x^2}}$$

$$\arg F = \arctan x.$$

The locus of F is a circle of radius $1/2$ centered at $z = 1/2$.

1.3.5 (a) $5e^{j53.1°}$
      (b) $7\sqrt{2}\,e^{j45°}$
      (c) $5e^{-j53.1°}$
      (d) $20.5e^{j11.3°}$
      (e) $e^{\cos 1}\,e^{j\sin 1}.$

1.3.11 (b) $1/50.$

1.3.15 *Hint:* use Euler's formula.

CHAPTER 2

2.2.3 $2^{23}(2^8 - 1) + 1.$

2.2.4 (a) $+1$
      (b) $-(1 + 16^{-1} + \ldots + 16^{-5})$

    (c) $1/128$
    (d) 16^{-65}
    (e) 16^{62}
    (f) Not normalized; -0 if legal.

2.2.8 Divide by 2, truncate fraction.

2.2.9 Multiply by 2, modulo 2^{n-1}

2.3.1 (a) $1/\sqrt{2}$
    (b) 1
    (c) $1/\sqrt{3}$.

2.3.3 $Q/4$.

2.4.1 The sampling frequency should be at least 20,000 samples/second.

2.4.4 rms error $= 1.13 \times 10^{-3}$.

2.4.5 (a) 1000 Hz
    (b)   500 Hz
    (c)   500 Hz
    (d) $\pi/5$ radians/sample interval
    (e) rms error $= 2.9 \times 10^{-3}$.

CHAPTER 3

3.1.1 (a) No, yes
    (b) Yes, yes
    (c) Yes, no
    (d) No, yes
    (e) Yes, no
    (f) No, yes
    (g) Yes, no
    (h) No, no
    (i) Yes, yes
    (j) Yes, no
    (k) No, yes
    (l) No, yes.

3.2.1

   (a) Phase $= -\omega$; amplitude $= \cos^2 (\omega/2)$.

   (b) Phase $= -\omega + \pi$; amplitude $= \sin^2 (\omega/2)$.

   (c) Phase $= -\omega$ for $0 \leq \omega < \pi/2$; $-\omega + \pi$ for $\pi/2 < \omega \leq \pi$;
      amplitude $= |\cos \omega|$.

   (d) Phase $= 0$ for $0 \leq \omega < \pi/2$; π for $\pi/2 < \omega \leq \pi$;
      amplitude $= |\cos \omega|$.

   (e) Phase $= (-3/2)\omega$; amplitude $= \cos^3 (\omega/2)$.

   (f) Phase $= -4\omega$ when $\cos 4\omega > 0$; $-4\omega + \pi$ when $\cos 4\omega < 0$;
      amplitude $= 2|\cos 4\omega|$.

   (g) Phase $= -4\omega + \pi/2$ when $\sin 4\omega > 0$; $-4\omega + 3\pi/2$ when $\sin 4\omega < 0$;
      amplitude $= 2|\sin 4\omega|$.

3.2.3 (a) $0, \pi/4, \pi/2, 3\pi/4, \pi$ radians/sample interval;

       (b) $0, 1250, 2500, 3750, 5000$ Hz.

3.3.6

   (a) Phase $= \arctan [-\sin 2\omega / (2 + \cos 2\omega)]$;
      amplitude $= \sqrt{5 + 4 \cos 2\omega}$.

   (b) Phase $= -3\omega/2$; amplitude $= \cos \omega/2$.

   (c) Phase $= -\omega$; amplitude $= 1$.

   (d) Phase $= 0$ for $0 \leq \omega < \pi/2$; π for $\pi/2 < \omega \leq \pi$; amplitude $= |\cos \omega|$.

3.5.4 $H(z) = z^{-1}$ suffices.

3.5.6

   (a) $H(z) = [(1+z^{-1})/2]^2$; zeros at $z = -1, -1$; poles at $z = 0, 0$.

   (b) $H(z) = [(1-z^{-1})/2]^2$; zeros at $z = 1, 1$; poles at $z = 0, 0$.

   (c) $H(z) = (1+z^{-2})/2$; zeros at $z = \pm j$; poles at $z = 0, 0$.

   (d) $H(z) = (z+z^{-1})/2$; zeros at $z = \pm j$; pole at $z = 0$.

   (e) $H(z) = [(1+z^{-1})/2]^3$; zeros at $z = -1, -1, -1$; poles at $z = 0, 0, 0$.

   (f) $H(z) = 1 + z^{-8}$; zeros at $z = e^{j(2k+1)\pi/8}$, $k = 0, \ldots, 7$; eight poles at $z = 0$.

   (g) $H(z) = 1 - z^{-8}$; zeros at $z = e^{j2k\pi/8}$, $k = 0, \ldots, 7$; eight poles at $z = 0$.

3.5.7 *Hint:* use Eq. (*) in the second example problem in Section 1.3.

3.5.8 (a) $H(z) = (1-z^{-1})[1-2z^{-1}+2z^{-2}][1-2z^{-1}+5z^{-2}][1-2z^{-1}+10z^{-2}]$.

       (b) $H(z) = 1 + z^{-4}$.

CHAPTER 4

4.1.1 Not in general.

4.1.2 $X(k) = 1$ if $k = 0$
 0 otherwise.

4.2.1 $X^*(z) = e^{1/z}$.

4.2.2

$$\frac{X(k)}{k+1} \xrightarrow{Z} z \int_z^\infty z^{-2} X^*(z)\, dz.$$

4.3.1 (a) $Y(k) = $ 1 for $k = 0$
 -1 for $k = 11$
 0 otherwise.

 (b) $Y(k) = $ 1 for $k = 0$
 2 for $1 \le k \le 10$
 1 for $k = 11$
 0 otherwise.

 (c) $Y(k) = $ 1 for $k = 0$
 -1 for $k = 1$
 -1 for $k = 11$
 1 for $k = 12$
 0 otherwise.

4.3.2 (a) $Y(k) = $ 1 for $k = 0$
 0 otherwise.

 (b) $Y(k) = $ 1 for $k = 0$
 2 for $k > 0$

 (c) $Y(k) = $ 1 for $k = 0$
 -1 for $k = 1$
 0 otherwise.

4.4.1

 (a) $1/(1 - z^{-1})^2$; zeros at $z = 0, 0$; poles at $z = 1, 1$.

 (b) $1/(1 - cz^{-1}) + 1/(1 - c^{-1}z^{-1})$; zeros at $z = 0$, $(c + c^{-1})/2$; poles at $z = c, c^{-1}$.

 (c) $1/(1 + .5z^{-1})$; zero at $z = 0$; pole at $z = -.5$.

 (d) $cz^{-1}/(1 - cz^{-1})^2$; zero at $z = 0$; poles at $z = c, c$.

 (e) $(a \cos b\, z^{-1} - 2a^2 z^{-2} + a^3 \cos b\, z^{-3})/(1 - 2a \cos b\, z^{-1} + a^2 z^{-2})^2$; poles at $z = ae^{\pm jb}, ae^{\pm jb}$.

(f) $(z^{-1}+z^{-2})/(1-z^{-1})^3$; zeros at $z=0,-1$; poles at $z=1, 1, 1$.
(g) $1/(1+z^{-1})$; zero at $z=0$; pole at $z=-1$.

4.4.4

$$\sqrt{(1-a\cos b\cos \omega)^2 + (a\cos b\sin \omega)^2}/\sqrt{(1-2a\cos b\cos \omega+a^2\cos 2\omega)^2 + (2a\cos b\sin \omega-a^2\sin 2\omega)^2}.$$

4.4.5 $F(k)=1$; $|z|>1$
$F(k)=c^k$; $|z|>c$
$F(k)=k$; $|z|>1$
$F(k)=a^k\cos kb$; $|z|>|a|$
$F(k)=a^k\sin kb$; $|z|>|a|$.

4.5.1

	$k=0$	1	2	3	4
(a)	1	2	3	4	5
(b)	1	3	6	10	15
(c)	0	.3	.24	.117	.0336
(d)	1	4	12.5	37.75	113.375
(e)	1	1	.5	0	$-.25$

CHAPTER 5

5.1.1

	$k=0$	1	2	3	4
(a)	1	1	.75	.5	.3125
(b)	0	1	2.5	4.25	6.125
(c)	0	.707	1.353	1.384	.692

5.1.2

(a) $H(z)=1+.8z^{-1}$
(b) Amplitude $=\sqrt{1.64+1.6\cos \omega}$
Phase $=\arctan (-.8\sin \omega)/(1+.8\cos \omega)$
(c) At $\omega=0$, amplitude $=1.8$, phase $=0$
At $\omega=\pi$, amplitude $=.2$, phase $=0$.

5.1.3

(a) $H(z)=1/(1+.8z^{-1})$.
(b), (c) Amplitude $=$ reciprocal of 5.1.2
Phase $=$ negative of 5.1.2.

5.2.1 Yes, if we assume that the output is a phasor if the input is.

5.2.2

    (a) $z^{-1}/(1+.1z^{-1})$
    (b) $z^{-1}/(1+.1z^{-2})$
    (c) $1/(1+.1z^{-3})$
    (d) $.5/(1+2z^{-1}-z^{-2})$
    (e) $(1-z^{-1})/(1+.5z^{-1})$
    (f) $1/(1+.5z^{-1})$
    (g) $(1+.2z^{-1})/(1+.2z^{-1})=1$.

5.2.3 Output $= 1, -.1, 0, 0, 0, \ldots$
A zero of the signal at $z=1$ cancels a pole of the filter at $z=1$.

5.2.4 Amplitude $= 1/\sqrt{1.25-\cos\omega}$
Phase $= -\arctan(.5\sin\omega)/(1-.5\cos\omega)$.

5.2.6 $1, 1.707, 1.957, 1.957$.

5.2.8 (a) $A=0, B=.81$
    (b) At $\omega=0$, amplitude $=1/1.81$, phase $=0$
         At $\omega=\pi/2$, amplitude $=1/.19$, phase $=0$
         At $\omega=\pi$, amplitude $=1/1.81$, phase $=0$
    (d) 250 Hz.

5.2.10 (a) $\frac{1}{2}\dfrac{1+z^{-1}}{1-z^{-1}}$
      (b) $|\frac{1}{2}\cot\omega/2|$
      (c) $-90°$.

5.3.1 (a) $(-1)^{k/2}, k$ even; $(-1)^{(k-1)/2}, k$ odd
    (b) $1.25(.5)^k-.25(.1)^k$
    (c) $(1/3)[(-1)^k+2\cos k\pi/3]$
    (d) $0, k$ even; $(-1)^{(k-1)/2}, k$ odd.

5.3.2 $(1/3)[-(1/2)^k+3+(-1)^k]$.

5.3.3 (a) $1/(1-z^{-n})$; (b) 4th order zero at $z=0$, poles at $z=+1,-1$,
      $+j,-j$.

5.3.4 (a) $z^{-1}/(1-z^{-1}-z^{-2})$; (b) zero at $z=0$, poles at $z=.5\pm.5\sqrt{5}$
      (real);
      (c) $((.5)^k/\sqrt{5})[(1+\sqrt{5})^k-(1-\sqrt{5})^k]$.

5.3.5 $[2 \cdot 2^k + (-1)^k]/3.$

5.3.7 (a) $Y^*(z) = 1/(1 - az^{-1})(1 - bz^{-1})$

(b) $Y^*(z) = z \left[\dfrac{a/(a-b)}{z-a} + \dfrac{b/(b-a)}{z-b} \right]$

(c) $Y(k) = [a^{k+1} - b^{k+1}]/(a-b)$
(d) $Y(k) = (k+1) b^k.$

5.3.8 (a) $X(k) = (.5)^k \cos k\,\pi/2.$

5.3.9 (a) $H^*(z) = 1/(1 - .25z^{-2})$
(b) Zeros at $z = 0, 0$; poles at $z = .5,\ -.5$; stable
(c) $H(k) = 0, k$ odd; $(.5)^k, k$ even.
(e) At $\omega = 0$, amplitude $= 1.33$
 At $\omega = \pi/2$, amplitude $= 0.8$
 At $\omega = \pi$, amplitude $= 1.33.$

5.3.10 $X(k) = 3/4 + k/2 + (-1)^k/4.$
The first few sample values are: $X(0) = 1;\ X(1) = 1;\ X(2) = 2;$
$X(3) = 2;\ X(4) = 3.$

CHAPTER 6

6.1.1 $F^*(n) = 0 \quad n \neq k$
  $= N \quad n = k.$

6.1.2 $F^*(n) = e^{j(x-n)(N-1)\pi/N} [\sin(x-n)\pi/\sin(x-n)\pi/N].$

6.2.1 *Hint:* prove that $Q(n)/n = \log_2 n.$

6.3.4 Original sequence: F_0, F_1, F_2, F_3
After bit reversal: F_0, F_2, F_1, F_3

After one merge: $F_0 + F_2, F_0 - F_2, F_1 + F_3, F_1 - F_3$
After two merges: $F_0 + F_2 + F_1 + F_3 \quad = F_0{}^*$
  $F_0 - F_2 - jF_1 + jF_3 = F_1{}^*$
  $F_0 + F_2 - F_1 - F_3 \quad = F_2{}^*$
  $F_0 - F_2 + jF_1 - jF_3 = F_3{}^*.$

CHAPTER 7

7.1.2 $n(n-1)/2$, where the graph has n nodes.

7.1.3 $n/2$ for n even; $(n+1)/2$ for n odd; where the graph has n nodes.

7.2.1 $2N \log N$.

7.2.4 *Hint:* use the result of Exercise 7.2.2.

7.2.7 The converse is not necessarily true.

7.2.10 n.

7.3.4 Let $n =$ number of nodes, then the branch-list method can require at most $n(n-1)$ locations; the adjacency-list method at most n^2 locations; and the adjacency-matrix method at most n^2 locations, each of which need only store 1 bit. This assumes no self-loops or multiple branches.

7.3.7 The degree of a node is n; there are $n2^{n-1}$ branches.

7.3.8 Yes.

7.3.9 Yes.

7.6.3 $2^{n+1}-1$ nodes, $2^{n+1}-2$ branches.

CHAPTER 8

8.1.2 One node connected to each of the remaining $n-1$ nodes by exactly one branch; a "star." There are n different such trees, one for each central node.

8.3.2 The last encountered tied distance would be retained, instead of the first encountered.

8.5.3 This is a sequence of moves that visits every square of the chessboard exactly once and returns to its starting square.

CHAPTER 9

9.1.3 The distance array method requires n^2 locations for the adjacency matrix, plus n^2 locations for the distance matrix (unless the distance matrix is made to serve as the adjacency matrix). The adjacency-list method requires at most n^2 locations for the adjacency list, plus $n(n-1)$ locations for the adjacency-distance list.

9.1.4 Store element in row i, column j, $j>i$, in position $(j-i) + (i-1)N - i(i-1)/2$.

9.2.2 The path is $A-F-D-C-M-L-B$ with 6 branches.

9.2.3 The shortest path is Philadelphia-Camden-Bordentown-Hightstown-Exit 9-New York, for a total of 110 minutes.

9.3.3 Possibly.

9.3.4 No.

CHAPTER 10

10.1.1

Figure 10.1.1 $B=\{(1,2), (2,1), (2,3), (3,2), (3,4), (3,5), (4,5),$
$(4,6), (4,8), (5,2), (5,6), (5,7), (6,7), (7,1),$
$(7,4), (8,1), (8,7)\}$

Figure 10.1.2 $B=\{(1,2), (3,1), (3,5), (4,1), (4,5), (5,2), (6,2),$
$(6,3), (6,4)\}$

Figure 10.1.3 $B=\{(1,1), (1,2), (1,3), (1,4), (1,5), (1,6), (2,2),$
$(2,4), (2,6), (3,3), (3,6), (4,4), (5,5), (6,6)\}.$

10.2.1 Symmetric: no
Reflexive: no
Transitive: no (if we interpret parent so as not to include grandparent).

10.2.2 Symmetric: yes

Reflexive: no (although we could define this relation to be reflexive if we wished)

Transitive: yes.

10.3.1 Store the BEFORE (I,J) list in row I; NUMBER (I) will contain the number of nodes that precede I.

10.5.1 Flow of water in a pipeline network with leaks at the nodes.

10.6.1 5.

10.6.2 10.

10.6.3 2.

10.12.5 Maximum flow $= 14$.

10.13.1 Man 1 to machine 8, 2 to 11, 3 to 10, 4 to 7, 5 to 12, 6 to 9; with all 6 men busy.

CHAPTER 11

11.2.1 *Hint:* resistor R_1 is irrelevant and can be crossed out.

11.2.2 *Hint:* convert the $V_0 - R_2$ series combination to a current source, combine parallel resistors, and convert back to voltage source form.

11.3.2

V_1	V_2	V_3	
5	0	-3	$= -4$
0	15	-7	$= 4$
-3	-7	10	$= 1$

11.4.1 $V_1 = -55/74$; $V_2 = 23/74$; $V_3 = 7/74$.

11.7.1

V_1	V_2	V_3	
$G_{01} + 1/j\omega L_{12}$	$-1/j\omega L_{12}$	0	$= I_{01}$
$-1/j\omega L_{12}$	$1/j\omega L_{12} + j\omega C_{23}$	$-j\omega C_{23}$	$= 0$
0	$-j\omega C_{23}$	$j\omega C_{23} + G_{03}$	$= 0$

11.7.3 (a) $-i(t) + Cdv/dt + v/R = 0$

(b) $\dfrac{V}{I} = \dfrac{1}{j\omega C + 1/R}$

(d) Low-pass.

11.8.1

$$\frac{V_3}{I_{01}} = \frac{1/G_{01}G_{03}}{1/G_{01} + 1/G_{03} + j\omega L_{12} + 1/j\omega C_{23}}.$$

INDEX